MEDDELANDEN FRÅN LUNDS UNIVERSITETS GEOGRAFISKA INSTITUTION
AVHANDLINGAR XCIX

Census from Heaven?

Population Estimates with Remote Sensing Techniques

Mikael Stern

1985
THE UNIVERSITY OF LUND, SWEDEN
DEPARTMENT OF GEOGRAPHY

Author's address:

Department of Physical Geography
University of Lund
Sölvegatan 13
S-223 62 Lund
Sweden

ISBN 91-7222-989-6

Wallin & Dalholm Boktryckeri AB, Lund 1985

contents

preface

My interest in remote sensing in semi-arid areas began with a pilot study on desertification in Tunisia in 1979, together with Dr Ulf Helldén. Later the same year the work with a doctoral thesis started at the Department of Physical Geography, University of Lund.

Financial support has been received mainly from the Swedish Agency for Research Cooperation with Developing Countries (SAREC) and the Swedish Board for Space Activities (DFR).

I am specially grateful to two persons for their support of my endeavour, my supervisor, Ulf Helldén, and my colleague, Lennart Olsson, who have continuously encouraged me, discussed my problems, and initiated new approaches.

Of course, all my colleagues at the Department are remembered at this moment, but they cannot all be mentioned here. However, I cannot omit Rezsö Laszlo, Ulrik Mårtensson, Eva Ahlchrona, Katarina Olsson, Karin Hall, Stefan Pinzke, Bengt Paulsson and Petter Pilesjö, who in different ways have made this dissertation possible.

The field work in the Sudan could not have been carried out without the support from Department of Geography, University of Khartoum, especially Mustafa Khogali, Siddig Awadallah.

I would also like to thank the Satellite Image Corporation and its helpful personnel, for letting me use their facilities.

Bengt Oreström should be remembered for erasing my linguistic errors.

Finally I must thank my family for their patience, and my friends for moral disintegration in times of exaggerated philosophical perspicacity.

abstract

The slightly ironic title seems to suggest that future censuses will be given to us by e.g. orbiting satellites. However, the question-mark implies a doubt, and the title was chosen to attract attention to the prospective possibilities to support today's traditional censuses by modern remote sensing techniques.

The study was initiated from a specific need for geographically accurate population data when investigating desertification processes in the Kordofan Province in the Sudan.

The basic idea was founded on the assumptions that people live in villages and that the sizes of the villages are correlated to the number of inhabitants. Thus, the investigation demanded data on population and settlement sizes. The former were collected through field work in Kordofan, whereas the latter were obtained from Landsat MSS data.

The bulk of the investigation was to suggest appropriate means to identify villages in the imagery, automatically or manually, and to compare these methods. Man convincingly outdid Machine, although the automatic methods were combined of spectral, textural and morphometrical discrimination processes.

Unfortunately, due to deficiencies of reliable ground truth, the high correlation between settlement size and population could not be statistically proved, but nor could it be denied.

1 introduction

'In those days Caesar Augustus issued a decree that a census should be taken of the entire Roman World. This was the first census that took place while Qurinius was governor of Syria. And everyone went to his own town to register' (Luke 2:1-3).

This quotation from the Holy Book not only depicts what is regarded as the first census ever, but also intimates the immediate problems arising in conjunction with a census: How count the population?

The classical enumeration dilemma is whether to count people by 'de facto' residence (the actual place at the census moment) or by 'de jure' residence (the legal place). Obviously, Augustus solved the problem by moving people back to their homes and making the actual place equal to the legal one.

Related counting problems are those of coverage and accuracy, not only due to ignorance, as illustrated in a comic strip (Fig 1.1), but relying on many parameters further discussed below.

The rather speculative and slightly ironic title 'Census from Heaven' seems to suggest that future censuses will be given to us by e.g. orbiting satellites collecting all needed demographic data, even fertility, mortality and health. However, the question-mark finishing the title implies a doubt. The title is chosen to attract attention to the prospective possibility to support today's traditional censuses by modern remote sensing techniques.

Aim of the study

To understand the disposition of the present investigation it must be remembered that the study originated from the need to solve a specific problem. During desertification studies at the Department of Geography, University of Lund, circumstantial spatial data on rural population density were lacking. The

Figure 1.1 B.C. by Johnny Hart.

main goal of the study is to design tools to avoid such deficiencies in a specific semi-arid area, the Kordofan Province in the Sudan (Fig 1.2). However, in the course of the methodology development four major utilities arose:

pro primo - support traditional censuses with the maps or sampling frames;

pro secondo - increase the spatial accuracy of censuses;

pro tertia - supplement traditional censuses with intercensal counts;

pro quarto - make rough, but fast counts in areas struck by e.g. drought.

These utilities are discussed in this introductory chapter, but the emphasis is put on the relations desertification - population - population studies - census, as well as how and why the proposed methods may support a traditional census.

1.1 Desertification studies

Based on earlier desertification studies and on discussions with researchers from Sweden, Tanzania, Kenya, Sudan and Tunisia a research project was launched in 1981 at the Department of Physical Geography, University of Lund, in co-operation with the Institute of Environmental Studies at the University of Khartoum. Financial support has been received from the Swedish Agency for Research Cooperation with Developing Countries (SAREC) and the Swedish Board for Space Activities (DFR).

The project was formulated for development, test and application of environmental monitoring methods for regional studies of desertification and its control (Rapp & Helldén 1979 and Helldén 1980, 1981). The recent investigations are mainly concentrated to the Central Sudan, where the objectives were to define relationships between various physical-biological parameters, and to model their variations with socio-economic variables.

'Desertification' expresses the arising of desert-like conditions in earlier unaffected areas. Most researchers postulate desertification, partly or wholly, to be ascribed to human over-exploitation of the environment. The effect is depicted as declining productivity in semi-arid areas (remote from the desert), rather than sand dunes tumbling from the desert erg to ravage inhabited land.

A commonly cited definition of desertification is the one by Dregne (1976):

> 'the impoverishment or arid, semi-arid and some sub-humid ecosystems by the combined impact of man's activities and drought. It is the process of change in these ecosystems that can be measured by the reduced productivity of desirable plants, alte-

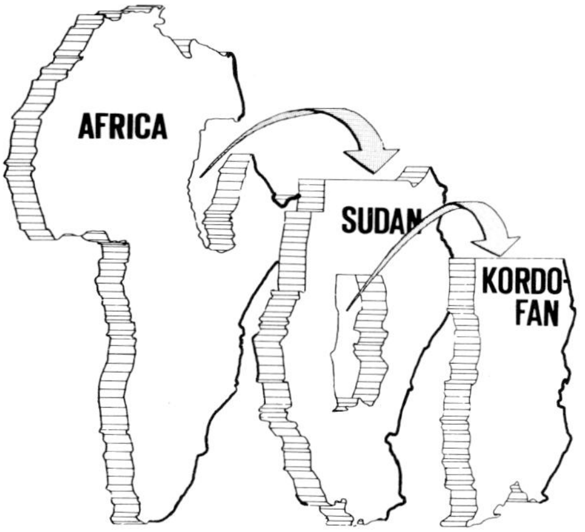

Figure 1.2 Kordofan Province, the Sudan (from Olsson & Stern 1981).

ration in the biomass and the diversity of the micro and macro flora and fauna, accelerated soil deterioration, and increased hazards for human occupancy'.

1.2 Desertification and human activities

Many processes (natural or social) related to desertification are interactions of cause and effect, e.g. migration flows, population decline or growth may be a result as well as an agent of desertification.

Probably 600-700 million people are living in regions menaced by desertification (United Nations 1977), whereof 78.5 million are living in areas currently becoming desertified (Novikoff 1983).

Livelihood systems

Land use in African drylands comprises three types of livelihood systems: animal-based, agriculture-based or urban-based. Often, however, these systems are mixed.

Animal-based

The animal-based systems herd cattle, sheep, goats or camels both nomadic and sedentary. Although the nomads utilize enormous areas for their livestock, the herds in more sedentary pastoral systems are much larger.

When the animals intensively graze grass and forbs, the barren ground will, temporarily, be exposed to wind and water erosion. Intensive browsing of trees and shrubs may change the micro environment (e.g. climate) for the flora and fauna.

The most favourable grazing species are the first to disappear from the savanna, less favourable plants decline in frequency more slowly, and in the environmental struggle the unpalatable species will flourish in overgrazed areas. Hence, a severely degraded area could be densely vegetated, but with undesirable plants (so-called green desertification). In the Sudan large areas are covered by such species as, for example, Usher (Calotropis procera), Marakh (Leptadenia pyrotechnica) or Sene Sene (Cassia senna or C. acutifolia).

When the large herds concentrate around wells and the fringes of villages, the trampling by hooves are even more devastating than the grazing.

Agricultural-based

The agricultural-based livelihood systems have two forms: rainfed and irrigated, and these suffer from different kinds of desertification. The special problem of irrigated cultivations are salinization and water logging, which deteriorate productivity.

The rainfed fields of the Sudan lie often barren from the harvest in October until the first millet straws crop up in June. For eight months essential nutrients are carried away with strong winds or by rare heavy showers. In times of declining yields the peasant is forced to curtail the fallow periods, which accelerate the impoverishment of the soil. Or he is forced to cultivate larger areas (if possible), perhaps further north towards more hazardous environment.

To most dryland populations wood is the main resource for fuel and construction purposes. When the sources in the neighbourhood are exhausted the wood is collected farther away, to increasing costs or efforts for the consumer. Wood gathering may lead to the same problems as intensive browsing of trees and shrubs.

Urban-based

Due to the need of water and shelter, aridity stimulates nucleated settlements. In times of environmental or political interferences the population seems to cluster even more. During the Sahelian drought in the early 1970's the urban population growth rate increased from 10% per annum to the double (United Nations 1977).

Desertification is an interference that may cause urbanization, but on the other hand urbanization is an aggressive factor of desertification. Increasing needs for energy resources, disposal of waste water etc, accelerate the stress on the urban hinterland.

The social consequences in an area struck by desertification is multifold. Apart from elusive psychological effects, like quality of life and emotional stress, there are a number of demographic indicators of desertification, for example of sedentarization, diversification or abandonment of settlements, or change of livelihood system (Reining 1978).

Needs for demographic data

The combat of desertification is much more a political problem than a technical one. But researchers have to serve the politicians with the base material for their decisions and actions.

The United Nations suggest in the report 'Desertification: Its causes and consequences' (United Nations 1977) that 'the assessment of physical conditions should be accompanied by efforts to obtain a more precise understanding of the state of dry-land peoples. Surveys should be undertaken, perhaps through a strengthening of census services and techniques, of their demographic characteristics, of the state of their health, and of their social and economic circumstances'.

The International Bank for Reconstruction and Development (IBRD 1975) pointed out that the major population characteristics ought to be of the highest priority in the general statistical policy in the Sudan. In addition to the population census, the demographic variables should be collected through regular sample surveys.

1.3 Census data

The general goal of a population census is to come out with a reasonable accurate count for a specific date. Since the census is the primary source of data on population dynamics, it is essential for all governmental planning regarding, e.g. population pressure or migration trends (Department of Statistics 1980a).

In addition to the the total count, data are also collected on demographic characteristics, e.g. fertility, mortality, education, marriage, household formation, economic activity, health, food and other consumption needs (Department of Statistics 1980a).

Census accuracy

A population census in the Sudan is a gigantic operation. Sudan censuses have been carried out three times, 1955/56, 1973 and 1983, of which the first one has been criticized for its low accuracy (see e.g. Henin 1968, Demeny 1968 and Som 1973), and the results of the second one has been withdrawn for the same reason. The 1983 census has not yet been analysed.

The lack of experience, of trained manpower and of economical resources make a census in Sudan awkward. Sudan is a vast area, and with the sparse population it is difficult to get a 100 % coverage. Also the multitude of languages spoken by the various tribe groups, the illiteracy and the ignorance decrease the accuracy of the data (Department of Statistics 1980a).

Bondestam (1973) states that data collected through field studies involves several difficulties: lack of appropriate field staff, limited transport facilities (especially in remote and thinly populated areas), heterogeneity of population, language difficulties, problems of reaching nomadic population, suspicion among local people, problems of cooperation with local authorities, etc.

Watson & Tippett (1975) mention that a complete enumeration census is exceedingly expensive, demands high levels of infrastructural development, use large numbers of skilled people, and will probably embody large amounts of bias.

Limitations on the use of census results in developing countries are slow processing, tabulation and publication of census material (Kpedekpo 1982). Note that results from the 1955/56 census in the Sudan was not published until about one decade later, and census material from the 1973 census was published in 1980. Watson & Tippett (1975) state that traditional total enumeration censuses are carried out without any objective testing of their efficiency.

Errors

Two main types of errors are common: errors of coverage (under-enumeration and over-enumeration) and errors of content. Coverage errors occur if villages, hamlets, households, persons, etc are not counted at all (Kpedekpo 1982).

Problems of over-counting or under-counting mobile populations arise because in order to count the total population, most censuses or enumerations are still carried out over a considerable period of time. Varying the census date from one count to the next also means enumerating very different numbers of seasonal workers (Kpedekpo 1982).

Errors of content may include faulty answers on questions. Respondents give ages which are systematically older than their real biological age. This is often done for prestige reasons or arises from African descriptive status - 'child', 'young man', 'elder' - which is then used to allocate a notional age to individuals (Kpedekpo 1982). Unlike developed countries, where age is a quantitative phenomenon, in most African countries age is a qualitative phenomenon (Gibril 1979).

Errors in reporting the type of economic activity have also been ascertained in censuses. Herdowners might be reluctant to report the number of heads in herd for taxation reasons, etc.

During the Second Population Census in the Sudan the enumerators should contact their alloted water holes and gather information on Sheikhs and household belongings. In some cases the enumerators were accompanied by the Sheikh himself, but in most cases the Sheikh was not found (Department of Statistics 1980a).

In the context of the extended family system and the seasonal migration, the methodology could result in omission or duplication. It is important to analyse the various stages and steps in the census operation to come out with possible reliability of the data. Consistency and convergency techniques as well as post-censal surveys are usually used to check the quality (Department of Statistics 1980a).

Sudan censuses

The 1955/56 census was done as a selected sampling and did not cover the entire population. The survey was carried out on a de jure basis, but de facto figures are available for some towns. The enumeration was scheduled for 12 months from July 1955, but it lasted for 14 months (Department of Statistics 1980a).

Being the first attempt, it did not collect very detailed data, but the focus on this count was only to facilitate the taking of a second census (Department of Statistics 1980a). But more than 17 years, until 1973, elapsed before the next census was launched.

The first figures from the 1973 census gave a total population of 12,265,028 against 10,262,536 of the 1955/56 enumeration. This would imply a growth rate of 1% per annum for the inter-censal period. Not only is this growth rate small, but for some segments of the population and some areas it was unacceptable.

The original figures were changed after the post-censal surveys (Department of Statistics 1980a), to the figures of Table I.1.

The problem of low growth rate may also be the opposite. Monsted & Walji (1978) pointed out that the growth rate between two censuses may be too high because of better census techniques and, hence, less under-enumeration in the later census.

From Table I.1 it may be concluded that the population increase is accelerating in all denoted provinces. For Darfur the acceleration is stable, while the other provinces have an increasing acceleration.

Geographical accuracy

To many researchers and planners, not only the demographic accuracy and the value of up-to-date information is of importance, but also the geographical accuracy of the data is essential. This is exemplified by Olsson (1985) when using precise

Table I.1 Corrected data on population, growth rate and density for some of the provinces of the Sudan. Compiled from the Second Population Census (Department of Statistics 1980a) and the Third Population Census (Department of Statistics 1984). The province Kordofan includes Southern as well as Northern Kordofan.

		Kordofan	Darfur	Khartoum	Sudan
Population	1955/56	1,764	1,331	427	10,015
(x 1000)	1973	2,416	2,309	777	13,811
	1983	3,093	3,094	1,802	20,564
Population density	1955/56	4.6	2.7	24.0	4.1
(inhabitants / sq km)	1973	5.5	4.2	52.1	5.6
	1983	7.1	5.6	120.9	8.3
Growth rate	1955/56 - 1973	1.8	3.1	3.5	1.9
(%)	1955/56 - 1983	2.0	3.1	5.3	2.6
	1973 - 1983	2.5	3.0	8.8	4.1

Table I.2 Livelihood system in Northern Kordofan 1983

Urban	235,348
Rural sedentary	1,077,540
Nomad	492,881
Total	1,805,769

geographical information for potential models, supply and demand models, etc.

The map for the 1983 census is a good example of bad geographical accuracy. The topographic map made in 1936, with some manually plotted corrections, was used as a base map. Fig 1.3 illustrates the villages on the census map over the study area in Kordofan, compared to villages in 1977 as interpreted from low altitude aerial photographs.

Several villages are omitted, some have mysteriously risen (probably "moved" from outside the photo coverage), and some of them are mapped as far as 6 kms off. There might be some minor changes from 1977 to 1983, but it must be remembered that 64 out of 65 villages in the study area in 1977 existed already in 1962, on exactly the same spot. The changes in the rural Kordofan are not swift.

Hence, the traditional census methods do not fulfill our needs for short-interval intercensal population estimates, and precise geographical information on population parameters. Similar problems with traditional census data in developing countries have been noted by e.g. Allen & Alemayehu (1975), Dayal & Khairzada (1976) and Anderson & Anderson (1973).

1.4 Remote sensing and population censuses

Aerial platforms

The use of aerial platforms for the collection of demographic data has been common for several decades. One unconventional method was used during the National Livestock Census for Sudan 1975/76 (Watson & Tippett 1975 and Watson et al 1976), where a capacious count was made in the form of a sampling from light aircraft, supported by ground investigations.

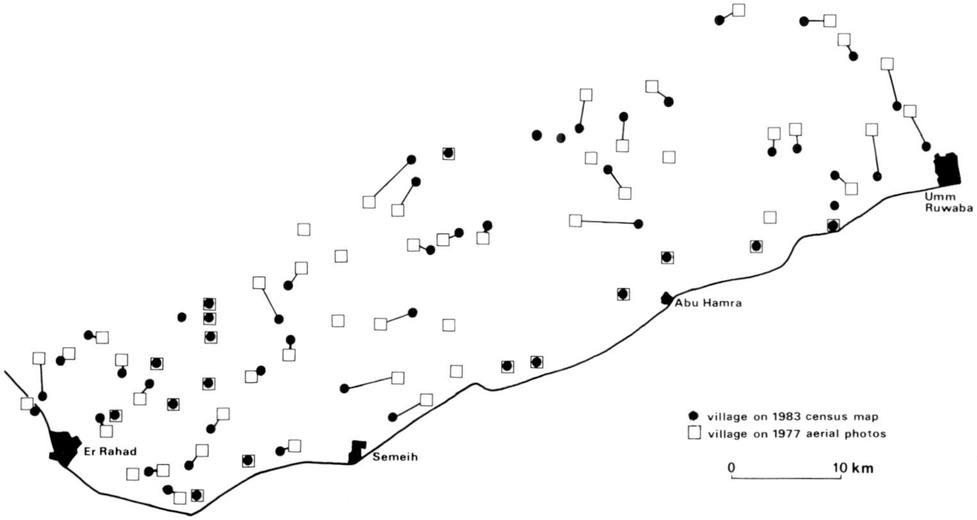

Figure 1.3 Location of censused villages (1983) compared to villages interpreted from aerial photographs (1977).

Table I.3 Examples of papers on population estimates, and variables to which they were related. *) Physical factors are here synonymous with a row of variables, e.g. area and highway distance to nearest larger urban neighbour. Methods explained by Holz et al (1969).

Variable	Data	Reference
Land use	Landsat	Morrison (1980)
Urban land use	Aerial photos	Kraus et al (1974)
		Olorunfemi (1984)
Settlement size	Landsat	Welch (1980)
	Aerial photos	Anderson & Anderson (1973)
	Radar	Harris (1985)
Physical factors*	Landsat	Ogrosky (1975)
Spectral density	Landsat	Iisaka & Hegedus (1982)
House count	Aerial photos	Lo & Chan (1980)
		Eyre et al (1970)

But more common for population estimates is the use of aerial photographs. Olorunfemi (1983) states that remote sensing can provide invaluable material to a census:

1) identification of settlements
2) settlement size and ordering
3) demarcation of enumeration areas
4) inter-censal population estimates

Olorunfemi (1984) pointed out that a population estimate from such methods lends itself to use in communities located in areas for which data on population are very scanty because of remoteness, political obfuscation, or insufficient resources to conduct frequent census enumeration.

Space platforms

Village identification from aerial photographs is an accurate method, but time-consuming and expensive, if the photographs are available at all. Identification from Landsat data, both manually and automatically, has been tested by several researchers. One example from Sahel is given by Reining & Egbert (1975).

Landsat MSS data was used by Reining (1980) to map population density and carrying capacity in the rural Upper Volta and Niger. Several researchers have also utilized remotely sensed data to study urbanization.

Jensen (1983) presented a good compilation of urban studies made with radar images.

Several authors have suggested that remote sensing from a satellite platform may support an enumeration census-type (Table I.3). The U.S. Census Bureau examined how Landsat data can be applied in demography tests in Bolivia and Kenya (NAS 1977). Clayton & Estes (1980) pointed out that the magnitude of discrepancy between imagery and census-derived counts at the tract level is relatively small.

Ethics

All researchers do not judge that remote sensing is a satisfactory tool in demography. In a recent debate (which elucidates the pros and cons in the contemporary discussions) in 'Photogrammetric Engineering and Remote Sensing' Brugioni (1983) first noted the low accuracy of the 1980 U.S. census, and also stated that the use of space-age technology can perform enumerations at a fraction of the cost of a traditional census.

Morgan (1984a) agreed with Brugionis argumentation and refered to population estimations he had made in Nigeria.

Morrow-Jones & Watkins (1984) partially agreed with Brugioni, but pointed out some of the technical problems (imagery, variable definitions), data problems and the uses of the information. They concluded that remote sensing may support a census, but it cannot be the principal survey method. They furthermore discussed the questions of ethics, which they felt are not often enough addressed in the field of remote sensing.

Also Sinclair (1984) raised the integrity argument against Brugioni: 'The outcry over invasion of privacy would be tremendous. The idea of spy-in-the-sky technology being used to count Americans would be a difficult pill for the citizenry to swallow, conjuring images of Big Brother that a novelist would envy'.

Brugioni replied that the methods advocated by him was not a surveillence of individuals: 'Indeed, there is no such capability from aerial reconnaissance'. The proposed method, he declared, is much more impersonal than the current census, which demands an individual's name and the answers to a number of very personal questions.

Morgan (1984b) stated that the use of aerial photography have the potential, in the long run, for being much more accurate than the traditional census figures. Paul (1984) concludes the debate by stating that this technology can only be used in conjunction with the traditional methods, and that social science applications of remote sensing technology are in general undervalued and underutilized.

Working with satellite data, which for civilian platforms will not exceed 10 m resolution during the next decade, there is no danger to tread on the personal integrity. However, one must be aware of the national integrity and carefully analyse who will profit from population estimates with satellites.

1.5 General approach of the study

Various approaches to the use of remotely sensed data for populations estimates have earlier been applied by other researchers. Table I.3 indicates some utilized variables correlated to population.

Forster et al (1984) summarized the major difficulties encountered using satellite data for studies of human settlements:

'... the urban landscape exhibits an extremely heterogenous surface cover, with considerable inter-pixel and intra-pixel change. This variation severely limits the standard computer techniques used to classify land use from spectral response and is primarily a function of the sensor resolution'.

The problems here exhibited by Forster et al, brought about the development of a number of special-made computer programs, which are only mentioned or briefly described in the successive chapters and hence the text hardly proportionate the efforts to construct them. Several of the used programs were created by Lennart Olsson and Stefan Pinzke, Remote Sensing Laboratory, Lund.

The quotation from Forster et al also illuminates the need for high resolution satellite data. In this investigation only Landsat MSS data were used, since they were the only data available, but the drawbacks and advantages of high-resolution data (Thematic Mapper and Spot data) are also discussed.

The current investigation was undertaken in five major steps:

1. Collection of ground truth, where population data on the village level are obtained in the field

or through population statistics (Chapter 10).

2. Identification of villages on remotely sensed data, which is the bulk of the present study. Chapters 4 through 8 discuss automatic methods, whereas Chapter 9 scrutinizes manual interpretation.

3. Measurements of village size and spectral density (Chapter 10).

4. Statistical correlation between number of inhabitants and measured village size (Chapter 10).

5. Mapping of population parameters. Final test reported in Chapter 11.

The general methodology was found (Stern 1981) to be useful in a given area in the Sudan, since all settlers live in compact villages large enough to be detected in remotely sensed data, all villages are similarly organized and all villages are situated on sandy soils in a rather open savanna.

The method development in steps 1-4 was carried out in a thoroughly investigated study area, while the final test was applied to an adjacent census district with less 'ground truth' information.

1.6 Laboratory resources

Fortunately, the Remote Sensing Laboratory in Lund is endowed with advanced hard-ware and soft-ware for computer processing of satellite data, as well as instruments for manual interpretation. During the present investigation was utilized:

- Dipix Aries II, an advanced image processor, built on a DEC PDP-11/73 and also comprising tape drive, disk memory, video refresh memory, colour monitor, graphics terminal, matrix printer and digitizing table. The system provides sophisticated programmes like image-to-image and image-to-grid registration, principal component analysis and maximum likelihood classification.

- Ebba, a more elementary, but easily handled, image processing system, based on a Metric-85 micro computer (CPU 62 Kbyte) and also comprises tape drive, video memory, TV-camera and colour monitor. Ebba was mainly used for simple parallelepiped classifications and statistical calculations.

- DIAPAC, an internally constructed 'Digital Image Analysis PACkage', applied to the Univac 1100/80 at the University of Lund Computer Centre. DIAPAC was originally constructed for elementary education purposes, but is also suitable for program development, because of the accessibility of graphical (UNIRAS) and statistical (NAG and Minitab) packages at the Computer Centre.

- Zoom Stereo Transfer Scope, which is an instrument for manual image interpretation, has the capacity to optically register images to maps, etc.

Some of the analyses have also been carried out at the Satellite Image Corporation, Kiruna, where two main systems have been used:

- IIS System 600, which is a system in the same range as Aries II, and is built on two VAX 11/750.

- Ebba II, which is a development of Ebba described above, but connected either on an IBM PC or a VAX.

A short test was also performed at a GOP-300 system with the courtesy of Context Vision, Linköping.

Available remotely sensed data of the study area:

- Landsat MSS data

Scene	187/51	1	January	1973
Scene	187/51	19	March	1979
Scene	187/51	29	September	1979
Scene	188/51	7	February	1973
Scene	188/51	9	November	1973
Scene	188/51	13	October	1979

- Aerial Photographs

1962	altitude	6700 m
1977	altitude	2800 m

2 settlement morphology

As introductory information to the analyses of rural settlements with remotely sensed data, this chapter treats the background to the appearance of settlement patterns, internal structure, shape, etc.

2.1 The study area

The study area for the pilot investigation (Stern 1981) of population estimates from Landsat data was chosen in the southern parts of Northern Kordofan.

The present sequel investigation was launched in an adjacent area, chosen entirely because of its outstanding coverage of rather recent (1977) aerial photographs. However, the study area is a representative part of Northern Kordofan.

Henceforth, the following definitions are used:

- study or test area: the entire area as demarcated in Fig 2.1;

- study or test sub-area: a part of the study area used for smaller tests.

Soils and cultivation

The area comprises one major land system, the Qoz zone, which is a vast area of stabilized parallel sand dunes, including some inter-dunes with hard red clayey soil (gardud). In some places the dunes are smoothed out to a homogeneous sand layer. Whether such an area should be called Qoz or not is disputed. The large wadi Khor Abu Habl, which borders on the study area, consists of black cracking clay (tin).

Although the Qoz is a very poor soil regarding nutrient status and water holding capacity, it is the most productive land type in Kordofan. Its loose structure facilitates cultivation and root penetration, and the sandy texture makes rainfall

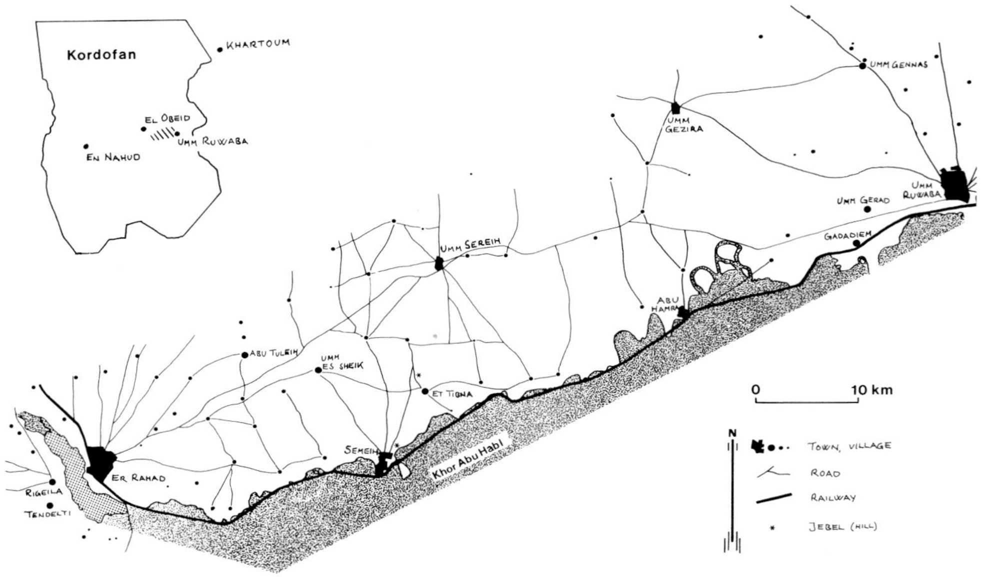

Figure 2.1 Map of the study area.

acceptance very effective.

The cultivation practised in the Qoz area is rain-fed agriculture with few exceptions. The water supply is mostly insufficient for large irrigated schemes, but some exist in Khor Abu Habl. The crops grown are mainly bulrush millet, sesame, water melon and ground nuts. The area is also a major production area for gum arabic (Acacia senegal), though the production has been fast declining due to, inter alia, deforestation.

The clay is heavy to cultivate without mechanized equipment, and during the rainy season these areas are often affected by flooding.

People and climate

The non-migratory tribe Gawama'a, which dominates in the study area, consists of at least 11 subtribes according to Watson et al (1976). The area is also inhabited by Ghodiat and Habbania (Barbour 1961). North of the study area Hamid is dominating, and to the west Hamar and Bederia prevail.

Watson et al (1976) point out that the nomadic Hawazma graziers, on their return north to the wet season grazings occupy fixed types of houses, and were therefore included in their population survey.

The climate of the study area is semi-arid, with a yearly mean precipitation between 350 and 400 mm, and the rainfall is concentrated to July, August and September.

More general information about the study area can be found in e.g. Barbour (1961), Born (1965) or Olsson (1985).

2.2 Settlement pattern

The geometric pattern of villages can be measured and divided into three categories, random, clustered and uniform pattern. The ultimate uniform pattern is the classical Christaller hexagons. The spatial settlement pattern might give important information on the allocation of the land and the land use.

A clustered pattern is supposed to be generated by contagious processes, i.e. one settlement attracts other settlement locations in its vicinity, e.g. because of a market or a well. A uniform pattern is caused by the opposite process, the competition for space between adjacent settlements (Olsson 1985).

The settlement pattern in the study area is assumed to be caused by the interaction between the two forces, competition for land and proximity to the main market (Olsson 1985).

Since the clay is hard to cultivate, the villages are preferably settled on the Qoz, close to the fields. Another reason for this location is the recurrent floodings of the clay plains.

The rule that all settlements are situated on Qoz land has very few exceptions in the study area. The only houses found on fine-grained soils are brickyards, houses around some large wells, or temporary dwellings.

Where the Qoz have discernible sand dunes and interdunes with clay or silt, the villages are localized at the crown of the dune. The dominating regular pattern entails a regular row of villages along longitudinal dunes and along khors and wadis.

2.3 Settlement size

The water supply rules the settlement size (el Bushra 1967), probably more than e.g. communications or markets do. Several types of water supply exist, e.g.:

Rahad, a natural accumulation of water in a wadi. In the study area there is only one rahad, and it has also named the close-by town, Er Rahad. This lake absorbs numerous herds during the dry season;

Hafir, an artificial surface reservoir. It can hold water all year around, and therefore encourages new settlers;

Well, (dug or bored) either hand-operated, perhaps with a shaduf (lever), or with a diesel-powered pump (donkay).

Tebeldi, also called baobab tree (Adansonia digitata), has a soft pith, which can be hollowed out and function as a water storage during the dry season (Fig 2.2).

Places with plenty of water attract more settlers, and they are even more numerous during the dry season. The number of inhabitants in villages with a poor water supply and dependence on tebeldis is strictly limited. A population increase in such a village gives rise to satellite villages in the neighbourhood.

The average size of a village is about 400 inhabitants, who dwell on an area of about 15 ha, which equals 35 Landsat pixels. However, towns like Umm Ruwaba, with 26,000 inhabitants (1983), and small two-pixel hamlets are found in the area.

2.4 Settlement shape

Most villages are of an insular type, circular or elliptic, and sometimes clustered around a mosque or a market place. For the measurements of settlement shape consult Chapter 7. The villages are usually referred to as nucleated, and this shape is stimulated by aridity, the need for water and shelter (Barbour 1961).

Surprisingly, there are no linear settlements along railways or main roads, though some of these villages have traces of slight elongation at the fringe.

Figure 2.2 The field sampling vehicle resting under the knobbly branches of a Tebeldi tree.

2.5 Settlement structure

The structure of the village is rarely conditioned by the water supply, but more by tribe affiliation, livelihood system or surrounding environment.

There are two main varying traits of the village interior structure: regularity and distribution. Most villages in Kordofan are characterized by irregularity in village plan and compound shape (Fig 2.3). Facts about traditional planning of indigenous villages are vague (Tewfik 1977).

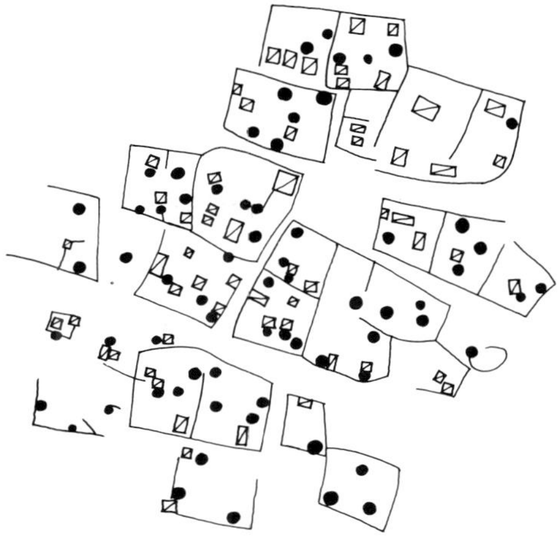

Figure 2.4 Regular structure.

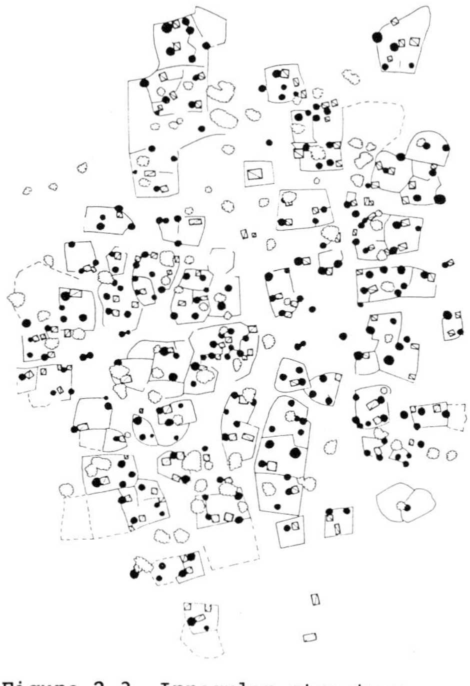

Figure 2.3 Irregular structure.

Figure 2.5 Scattered structure.

Some villages are more or less systematized into a grid. These are the most recently settled villages; they are influenced by modern cities (Born 1965) and are mainly found along the railway El Obeid - Kosti or close to towns (Fig 2.4). But there are villages all along the entire range from ostensible chaos to rectangularity, although most villages are organized like the example in Fig 2.3.

The degree of distribution depends mainly on livelihood system. According to Born (1965), a dense settlement pattern is a sign of peasant population, and a more scattered pattern (Fig 2.5) indicates stockraisers or perhaps settlers with a nomadic heritage.

Villages with animal herders often swell in circumference due to the occurence of zeribas, enclosures into which the animals are driven during the night (Fig 2.6).

Figure 2.6 Village with zeribas.

The circumference or the density of many villages oscillates according to season. Non-nomadic fluctuations depend mainly on water shortage. Extensive immigration strikes mostly urban areas, but according to Ibrahim (1978) increases the population of, for example, the village Mellif (in Darfur) from 8,000 during the wet season to 17,000 during dry season.

The internal village structure is not always homogeneous, not even in small villages, and the diversification applies to both regularity and distribution. In trading villages the compounds of the shopkeepers and craftsmen cling to the market, and the peasants live in other parts of the village in other kinds of dwellings.

Around newly built wells different tribes settle on different sides of the water yard (Born 1965). Foreign immigrants in the study area are mainly West African pilgrims, Fellata, on their way to Mecca. But many of them have settled, sometimes in separate villages and sometimes in a separate part of a town or a village. However, the settlement structure made by Fellata does not vary widely from traditional Gawama'a.

2.6 Compound structure

A typical compound in Northern Kordofan is fenced by a wall made of shrubs or millet stalks. Barbour (1961) explains that muslims regard the seclusion of women, and their protection from the eyes of other men, of greatest importance. Necessary is therefore the possession of a wall around a man's property, high enough to protect its inmates from observation by strangers (Fig 2.7).

A typical hut (tukl) is round, with a conical thatched roof, and is made of a wooden frame covered by grass and millet stalks (Fig 2.7). The number of huts depends on the size and wealth of the household, but the average is 3-4. Large huts are sleeping quarters or storage rooms, while a small hut is usually a kitchen.

The average household size in Kordofan 1973 was 4.94 persons (Department of Statistics 1980b). However, the standard deviation is very high (Fig 2.8). The average household size in 1983 was 5.94 for Northern Kordofan overall, and 5.85 for rural sedentary households in Northern Kordofan (Department of Statistics 1984). According to Watson et al (1976), the average in 1976 was 8.47 for Northern Kordofan and 8.29 for the Gawama'a tribe. The discrepant figures are scarcely occasioned by a rapid growth of household

Figure 2.7 Typical compound with tukl.

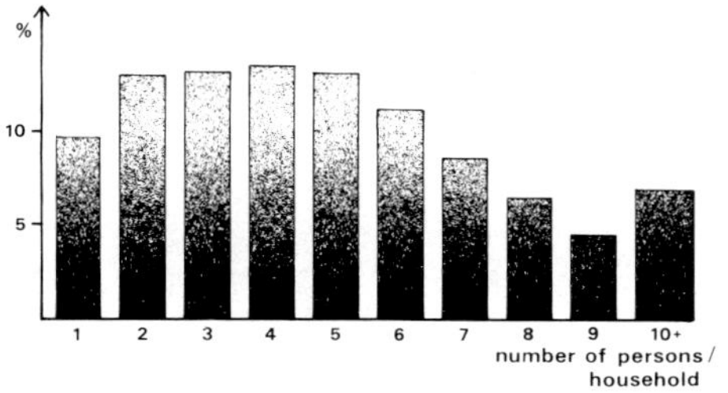

Figure 2.8 Number of persons per household in Northern Kordofan.

sizes, but by a difference in definition or data collection method.

In addition to the tukls are found flat-roofed sun-shelters (rakouba, see Fig 2.9), and in many villages also a couple of mudbrick houses accomodating a school, a hospital, a shop, or a police station.

The tukls, which are built within a

Figure 2.9 Rakouba.

Figure 2.10 Silhuette of shading trees in a village.

couple of days, are short-lived dwellings, and have to be improved or rebuilt after heavy rains. The average life time is about three years (Reeves & Frankenberger 1981). Sometimes the resettled compounds are erected in another place or in a different manner, which makes the internal structure of a village instable.

2.7 Village vegetation

A typical village silhuette is also characterized by its trees (Fig 2.10). The villagers keep the trees mainly for the shade and they are not used for fire-wood etc. Common species are heglig (Balanites aegyptica), haraz (Acacia albida), neem (Azadirachta indica), sayyal (Acacia tortilis) and sidr (ziziphus spina-christi). The tebeldi tree (Adanasonia digitata), which is used for water storage, is not common inside the village.

2.8 Village perimeter

From a remote platform the vision of a village also depends on its immediate surroundings. A typical village in the study area is encircled by a barren patch (10-500 m), called khatwat el-hilla ('the village step') in arabic. It is sometimes referred to as a village halo due to the light spot that appears on e.g. Landsat data (Fig 2.11).

The village halo is, according to most desertification researchers, to be explained by trampling and grazing by animals, and wood collection (see e.g. Babiker 1982 or

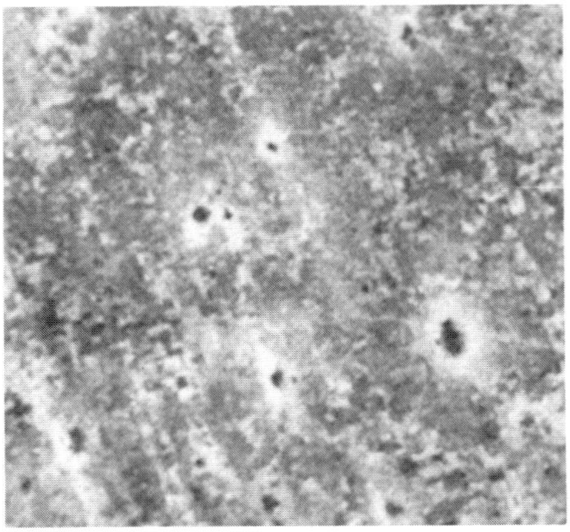

Figure 2.11 Typical villages with whitish halos.

Ibrahim 1978), and the villagers never attempt to cultivate this zone (Reeves & Frankenberger 1981).

In some villages the inhabitants claim that they 'want to keep their village clean' or that the local Health Service enjoined them to keep bushes away from the fringe of the village. Other reasons are the function as fire break, or remnants from old tribe wars, when the halo may have allowed a possibility to spot approaching dangers/enemies (Helldén 1984).

Helldén (1984) found no correlation between village size and size of village halo. Nor did he find a trend of increasing village halos, when he studied aerial photographs and satellite images from 1962, 1972 and 1979. The existing fluctuations can, however, reflect the rotational land use. His findings conflict with the ones of DECARP (1976), which proposes a continuously increasing process. However, Hellden found a relation between type of water supply in the village and the size of the halo.

Not all villages are totally encompassed by a village halo, sometimes cultivated fields and grazing land border the village. Sometimes the halo is endowed with a thin forage cover which is used to graze donkeys, goats and camels kept in the village (Reeves & Frankenberger 1981).

The rather whitish village halo may, after heavy showers have scudded the country, be transformed into a greenish grass halo, which even may gleam red on a Landsat FCC. Hence, the halo can be fertile and cannot, senso strictu, be called desertified.

The land beyond the village halo consists of cultivated fields (zira'a) or bushland dominated by ushar (Calotropis procera), marakh (Leptadenia pyrotechnica) or sene sene (Cassia senna), species flourishing in overexploited areas, since they are refused by the animals.

3 spectral properties of a village

The spectral response of an arid natural environment is the sum of the reflectance of the vegetation, the soil (or the bedrock) and their shadows. The human environment also includes different types of anthropogenic constructions.

To unravel why most villages in the study area are clearly visible on Landsat data it is necessary to thoroughly explore the spectral properties of its components.

3.1 Component coverage

When investigating the total spectral response of a village, the first problem to solve is the coverage of the village components. This distribution varies widely from village to village, but a case study of 5 typical Kordofan villages in the Er Rahad area is a suitable example.

The spatial distribution of the main components: ground, trees, huts, houses, sun-shelters and compound walls, in relation to the size of the total village inside its outer boundaries, was measured from aerial photographs (Table III.1).

Table III.1 Coverage of the village components of 5 villages in the Er Rahad region.

Component	Coverage
Trees	4 %
Huts	5 %
Houses or sun-shelters	3 %
Walls	3 %
Ground	85 %

The human dwellings cover apparently only about 11 %, and the trees about 4 % of the total village area. The remainder is spectrally heterogenous sand, where half of it is barren sand and the rest is sand covered by litter, ashes and excrement.

The amount of shadows in the village is dependent on sun angle at the moment of data acquisition and height of the objects. The shadows cover an area of maximum 20 % of the area of dwellings and trees for the used remotely sensed data. And since the shadows only make a considerable spectral difference when they fall on bare sand, the shadow affects only 1-2 % of the total area, and may be neglected.

Spectrally, everything except bare sand distinguishes the village from its immediate surroundings, the village halo. If bare sand within the village boundaries is deducted from the village area, the percentage of the village components is changed. The distinguishing parts of the village consists of 74 % polluted ground and 26 % dwellings and trees. Hence, what characterizes a village, as perceived from a remote nadir sensor, is much more the litter from its residents rather than their huts and houses.

3.2 Component reflectance

The reflectance of an object varies with the wave-length of the light. If the spectral properties of the village components should be closely compared with Landsat data, a hand-borne radiometer with filters corresponding to the four spectral bands is needed. Such advanced facilities were not available in the study area. However, spectral measurements were carried out on aerial pan-chromatic black-and-white photographs, which almost correspond to Landsat MSS 4 and MSS 5 (Fig 3.1).

The measurements were relative since no absolute calibration method was available. However, a rough calibration was carried out based on the normal reflectance of sand (according to Swain & Davis 1978 and Reeves et al 1975) and a roughly simulated black body.

Fig 3.2 depicts the theoretical reflectance curves for some urban phenomena, gathered from different literature sources.

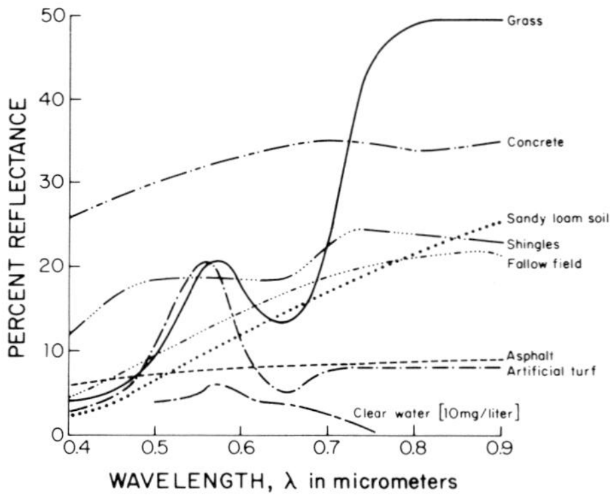

Figure 3.2 Theoretical reflectance curves for some urban phenomena (from Jensen 1983).

Table III.2 indicates the relative reflectance of the village components, and their approximate absolute values in the visible spectrum. The values should be regarded as having large confidence intervals, due to the variance of sand reflectance. However, the inter-relation between the values is quite reliable. Shadow and polluted sand have higher standard deviation than the others.

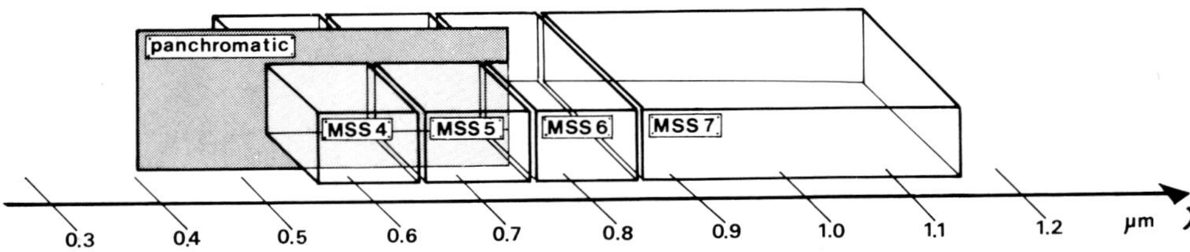

Figure 3.1 The spectral relation between pan-chromatic film and the four Landsat bands. (The figure is stylized, in reality the sensitivity of the sensors is uneven, and they interfere with each other.)

Table III.2 The approximate absolute values of reflectance in the visible spectrum for the village components.

Component	Reflectance
Sand	25 %
Polluted sand	17 %
Hut	17 %
Wall	16 %
House or sun-shelter	27 %
Tree	12 %
Shadow	10 %

Notable is that the flat-roofed sun-shelters (some of them may be brick houses) have a higher reflectance than the bare sand, and are therefore negative for the village distinction from its surroundings.

The average reflectance of the village was calculated to approximately 20 % (bare sand included) and 17 % (bare sand excluded) compared with 25 % in the halo engulfing the village. The average reflectance may proportionate the population density of the village since all objects (including the amount of polluted sand) are functions of human activities. The integrated optical density, IOD, of the village, i.e. the total reflectance, might thus reflect the total population of a village, and is further discussed in Chapter 10.

3.3 Mixels

Since the village consists of a row of objects with highly varying reflectivity, a very high variation in the Landsat data is anticipated. However, the village components described above are relative to the

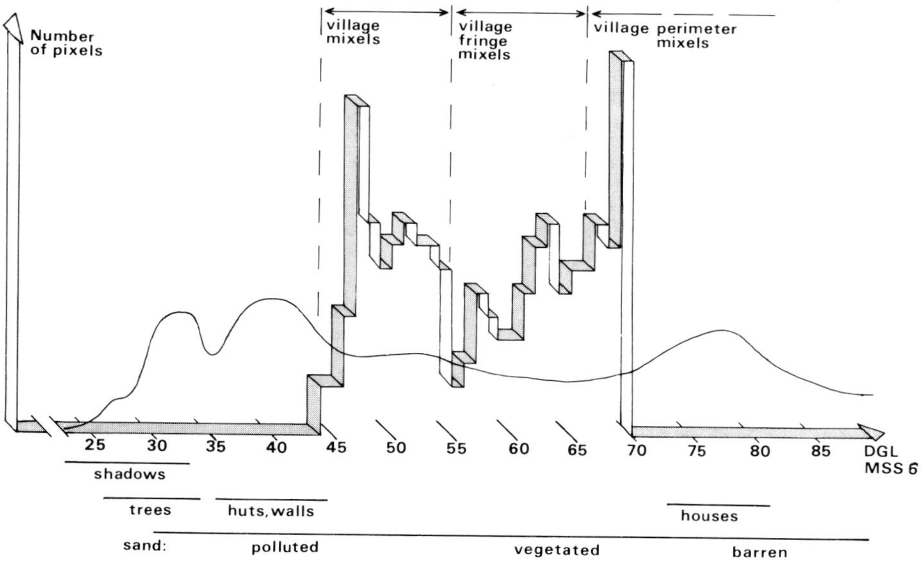

Figure 3.3 Frequency reflectance histogram of a typical village. The smooth curve indicates the 'true' reflectance. The 'camel bumps' to the left are effected by trees, shadows and huts, whereas the 'bump' to the right represents the houses and barren sand within the village (areas outside the village are not represented). The rugged histogram in the middle illustrates the MSS 6 DGL frequencies of the village, including the village halo. Note that the entire area of 'true' values is compressed into the interval called 'village mixels' in the MSS histogram.

MSS pixel very small. A straw hut, for example, covers about 10 sq m, a tree might be 25 sq m, whereas the instantaneous field of view, IFOV, for MSS data is more than one hundred times larger, approximately 4000 sq m.

It is easily comprehended that the same effect will appear for TM data, the IFOV of a TM pixel is approximately 900 sq m. Not until SPOT data is available will there be a sensor IFOV close to the magnitude of the objects in a Kordofan village. SPOT P-band sensor has an IFOV of ca 100 sq m. However, a high resolution is not necessarily beneficial for spectral or multi-spectral classification (see e.g. Jaakkola et al 1984).

The digital grey level, DGL, of a village pixel is the reflection from a jumble of objects, why such a mixed pixel is referred to as a mixel. Smedes et al (1975) mention that 25-30 % of the classification error is caused by mixture problems. Hyde & Vesper (1983) found in their test area that 43 % of all MSS pixels and 100 % of the urban pixels were mixed.

Probably all pixels in a village are mixels, which means that the high

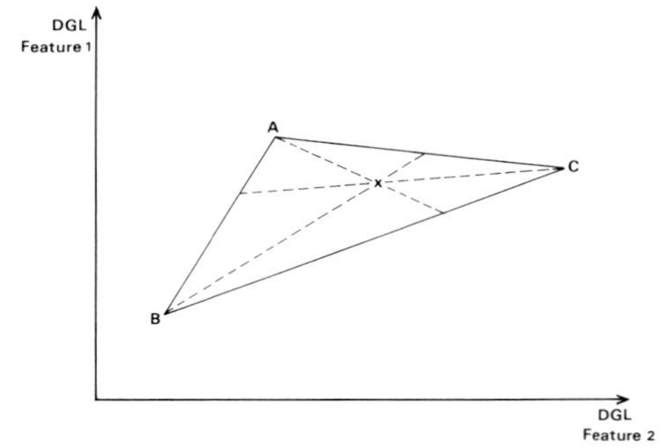

Figure 3.4 The mixel DGL value is composed of reflectance from the objects A, B and C (after Work & Gilmer 1976).

reflectance from e.g. bare sand or the low reflectance from e.g. a straw hut is neutralized and will hardly show up in a frequency histogram of a MSS band (Fig 3.3). Thus, the standard deviation is comparatively low although the reflectance intra-variation of the objects is high.

Looking at a 4-dimensional (one for each MSS band) scattergram this averaging effect will cluster the village pixels in the center of a 4-dimensional "polygon". In this hyperspace the "polygon corners" are the reflectance "co-ordinates" from the

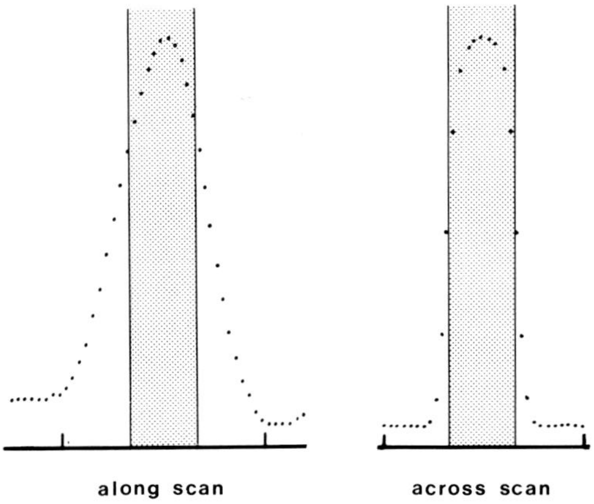

0.02	0.04	0.01
0.20	0.52	0.14
0.02	0.04	0.01

Figure 3.5 a) Landsat point spread function (after Dye 1975). b) Average convolution weights for a central pixel and its surrounding due to the point spread function (after Forster 1980).

spectrally pure village objects. From Fig 3.4, which explains in two dimensions, the 4-D effect is comprehendible.

3.4 Smoothing effects

Point spread function

Working on a pixel level it is essential to remember that the DGL of a pixel is derived not only from the target area (IFOV) on the ground, but also from the surrounding areas. This is due to the point spread functions (psf) of the atmosphere and the scanner.

According to Dye (1975), the two-dimensional psf of the scanner is the combined effects of the optical blur circle and the detector size in the across-scan direction and the pre-sampling low-pass filter in the along-scan direction (Fig 3.5). Similar effects are achieved by the atmospheric impact.

The effect on a Kordofan village is depicted in Fig 3.6. Note that the psf not only makes the village larger, but also extends the boundary pixels towards the centre of the village. The fictional psf used in Fig 3.6 is borrowed from Forster (1982). He also discussed filters to deconvolve the psf, which demands calculations using very precise information on the reflectance of the true target area, which was impossible in Kordofan.

Resampling

Resampling, employed when making geometric transformations of images, is a procedure which also affects the appearance of objects. There are three major interpolation methods (for a deeper discussion see e.g. Schowengerdt 1983):

- nearest neighbour;
- bilinear;
- cubic convolution.

The latter, which in some of its variants is the most credited, is by Pratt (1978) depicted as an approximation of the theoretically ideal interpolation function for imagery, namely

$$\frac{\sin x}{x}$$

Generally, it is assumed that nearest neighbour resampling, which is

a)

25	25	25	25	25	25	25	25	25	25	25	25	25
25	25	25	25	25	25	25	25	25	25	25	25	25
25	25	25	25	25	25	25	15	25	25	25	25	25
25	25	25	25	20	20	15	15	15	15	15	25	25
25	25	25	25	20	15	15	15	15	15	20	25	25
25	25	20	15	15	15	15	15	15	15	25	25	25
25	25	25	20	15	15	15	15	15	15	20	25	25
25	25	25	25	20	15	15	15	15	15	20	25	25
25	25	25	25	25	15	15	15	15	15	25	25	25
25	25	25	25	25	15	15	20	25	25	25	25	25
25	25	25	25	25	25	15	20	25	25	25	25	25
25	25	25	25	25	25	25	25	25	25	25	25	25
25	25	25	25	25	25	25	25	25	25	25	25	25

b)

25	25	25	25	25	25	25	25	25	25	25	25	25
25	25	25	25	25	25	25	24	25	25	25	25	25
25	25	25	25	25	24	22	19	23	24	24	25	25
25	25	25	24	21	19	17	15	16	16	19	24	25
25	25	24	23	19	16	15	15	15	16	20	24	25
25	24	21	17	16	15	15	15	15	17	23	25	25
25	25	24	20	16	15	15	15	15	16	21	24	25
25	25	25	24	20	16	15	15	15	16	21	24	25
25	25	25	25	22	17	15	15	16	18	23	25	25
25	25	25	25	23	17	16	20	23	24	25	25	25
25	25	25	25	25	22	18	21	24	25	25	25	25
25	25	25	25	25	25	24	24	25	25	25	25	25
25	25	25	25	25	25	25	25	25	25	25	25	25

Figure 3.6 a) "True" pixel values (15) and mixel values (20) of a fictional Kordofan village, and its halo pixels (25). b) The same village affected by the psf (convolved with weights from Fig 3.5b).

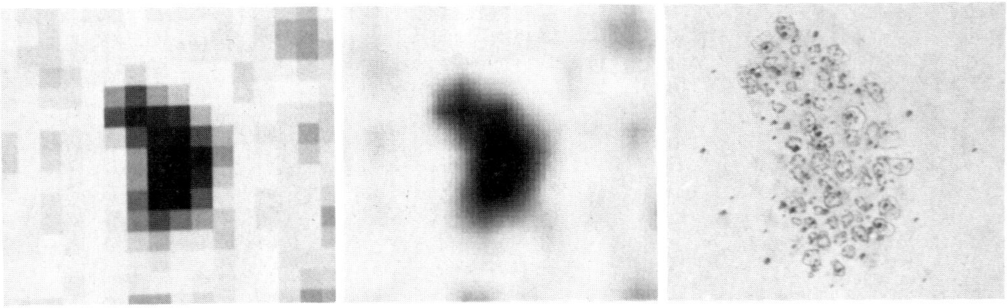

Figure 3.7 a) Landsat MSS image of a village, b) same village transformed with cubic convolution, c) same village on an aerial photograph.

the fastest method, creates an output image that is spectrally close to its origin, but with some adjoined geometric distortion due to repeating and omitting of pixel values. Cubic convolution resampling, on the other hand, changes the content of the image, but makes it closer to reality both spectrally and geometrically.

Verdin (1983), when studying lakes in Wisconsin, found that a cubic convolution resampling blurred the boundaries between lake and land pixels. It also created DGLs totally absent in the original images. Hence, for some studies non-corrected data may be valuable.

Forster & Trinder (1984) found, when classifying forest using raw MSS data and data resampled with cubic convolution, that the latter were more likely to be correct with respect to the ground.

Rescaling, which is often involved in resampling, does not degrade the images only if the pixel sizes of the images divide into each other evenly (Hyde & Vesper 1983).

Fig 3.7 illustrates the effects of cubic convolution applied on a village in Kordofan. The raw data of 3.7a is enlarged four times with cubic convolution and shown in 3.7b, while 3.7c is the same village as it appears on the aerial photograph.

3.5 Spectral response in Landsat MSS data

Keeping the mixel problem and the blurred data in mind we will now look further into the appearence of villages in Landsat data. In the study area there are 85 villages of

Table III.3 Reflectance means (and standard deviations) for the sample of villages from Fig 2.3, 2.5 and 3.9.

Data	typical	scattered	polluted	clean
March MSS4	28.7 (0.9)	33.9 (0.9)	31.0 (1.2)	30.3 (0.5)
March MSS5	48.1 (2.4)	63.1 (2.3)	51.8 (2.1)	54.4 (2.5)
March MSS6	55.9 (2.0)	68.4 (1.7)	59.0 (2.1)	63.7 (1.3)
March MSS7	48.2 (1.0)	60.3 (1.4)	50.7 (2.0)	54.9 (0.8)
September MSS4	26.7 (1.2)	27.9 (1.2)	28.0 (1.1)	27.1 (1.2)
September MSS5	38.7 (1.7)	42.2 (2.0)	43.3 (2.2)	41.9 (1.8)
September MSS6	56.8 (1.5)	59.8 (2.4)	57.8 (3.2)	59.0 (1.5)
September MSS7	48.4 (1.2)	55.1 (1.1)	47.8 (1.2)	50.0 (1.0)

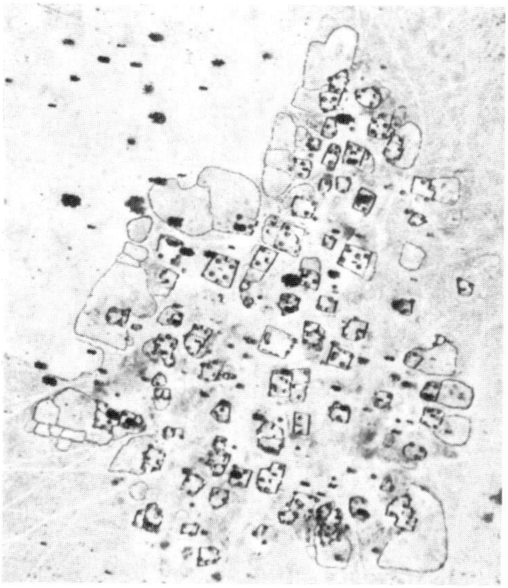

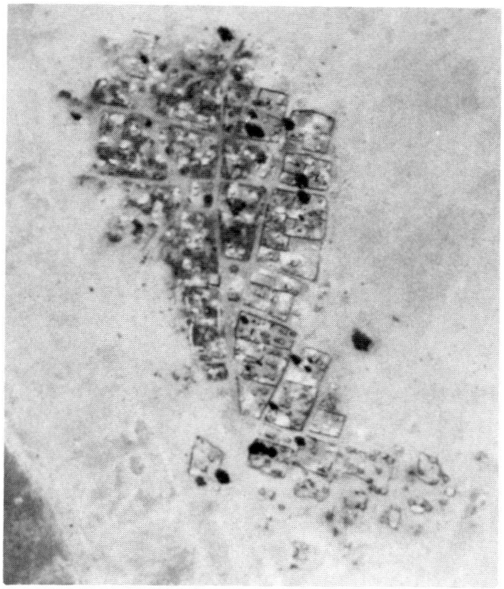

Figure 3.8 a) Unusually clean village. b) Unusually polluted village.

varying sizes and densities. Fig 3.9 shows an atypically clean and an atypically polluted village, and Table III.3 depicts statistics of tyical and atypical villages.

3.5 Conclusion

Cubic convolution resampling of the village image seems to be closer to the real village than the original image why further comparison tests will be employed in Chapter 7.

The villages seem to have a spectrally rather stable behaviour when changing internal structure, which is hopeful awaiting the coming classifications (Chapter 4 and 5), where the village response also will be compared to other land cover classes.

4 spectral discrimination of villages

This chapter deals with the possibilities to discriminate villages using their spectral properties only. Spectral clustering, which is the prevalent discrimination criterion, is recognized by the human eye as similar grey tone or colour.

Since data from two dates (8 bands) are available, much of the discussion concerns which information to use (feature selection). Passing the first rough classification tests, a sample of methods continues to further selection tests and evaluations.

Together with the procedures singled out in Chapters 5 through 7 the most preferable spectral method will proceed to the test of combined automatic methods (Chapter 8).

4.1 Feature selection

Definitions

A feature is 'a vector with components which are functions of the initial measurements pattern variables or some subsequence of the measurement vectors' (Colwell et al 1983).

Here, feature is the general concept of image information. Not only are the original Landsat MSS bands used, but also band combinations and band transformations, forming new features which hardly may be referred to as spectral bands. They all geometrically correspond to the original features, the MSS bands.

Other concepts henceforth used are class, training area and classification. Classes are thematic segments, as e.g. bare sand or millet cultivation, or spectrally homogeneous clusters. By training area is understood a representative sample of the class, while classification is an automatic class labelling of each pixel based on training area statistics.

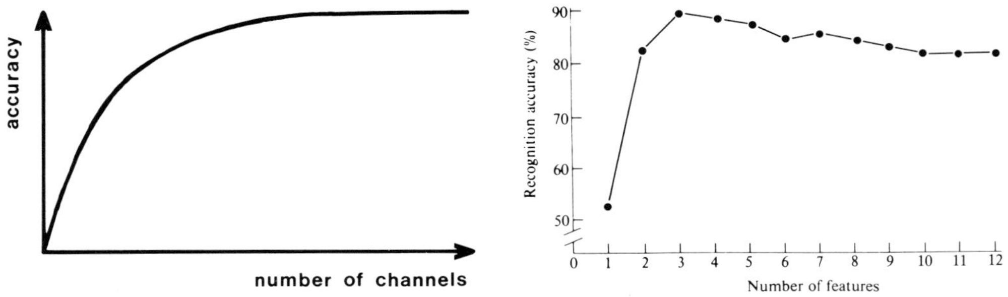

Figure 4.1 Relation between classification accuracy and number of features according to a) Bressanin & Erickson (1973) and b) Swain & Davies (1978).

Methods

In this chapter, features are selected contingent on their capability of discriminating villages from their surroundings.

Feature selection methods may be heuristic, where the selection is empirical, or based on radiometric theory or statistics. Heuristic methods are as predominant in applied remote sensing papers, as are the statistical methods in image processing essays.

Besides culling the most potent features, the selection process also involves limiting the number of features (feature reduction), which is a weighing between accuracy and costs. An increased number of features implies an improvement of the classification accuracy, but not necessarily a linear increase. Bressanin & Erickson (1973) propound theoretically an asymptotic relationship between accuracy and number of features (Fig 4.1a).

Swain & Davies (1978) are of a contrary opinion. They stated that the curve has a maximum for a fixed number of training samples (Fig 4.1b).

Principal component analysis

A common way to decrease the number of features is with principal component analysis, where a linear transformation of the original axes is performed. The so-called Karhuen-Loewe transformation is based on a diagonalization of the co-variance matrix. The canonical transformation involves training areas and maximizes the separability of the classes by maximizing the inter-class variability and minimizing the intra-class variability. For further explanation of the procedure, consult Schowengerdt (1983) or Curran (1985).

The first principal component includes the axis with the largest variance, a.s.o. The great majority of the variance may thus be included in fewer dimensions than the original set of features.

Most information from the four Landsat bands fit into the first three components, while the fourth component is almost entirely noise, and may be demolished.

Having 8 bands, interesting information might be present in more than three bands. The first four components were transferred, without further tests, to the combined automatic methods (Chapter 8).

4.2 Statistical separability

One useful measure when selecting features is separability, i.e. to what degree a village can be separated from other spectrally homogeneous classes. Separability is closely related to the inter-class variability. Another selection criterion is a low variation within the class or training area, called the intra-class variability, which is often both strong and natural.

Distance is the most commonly used separability measure. The spectral

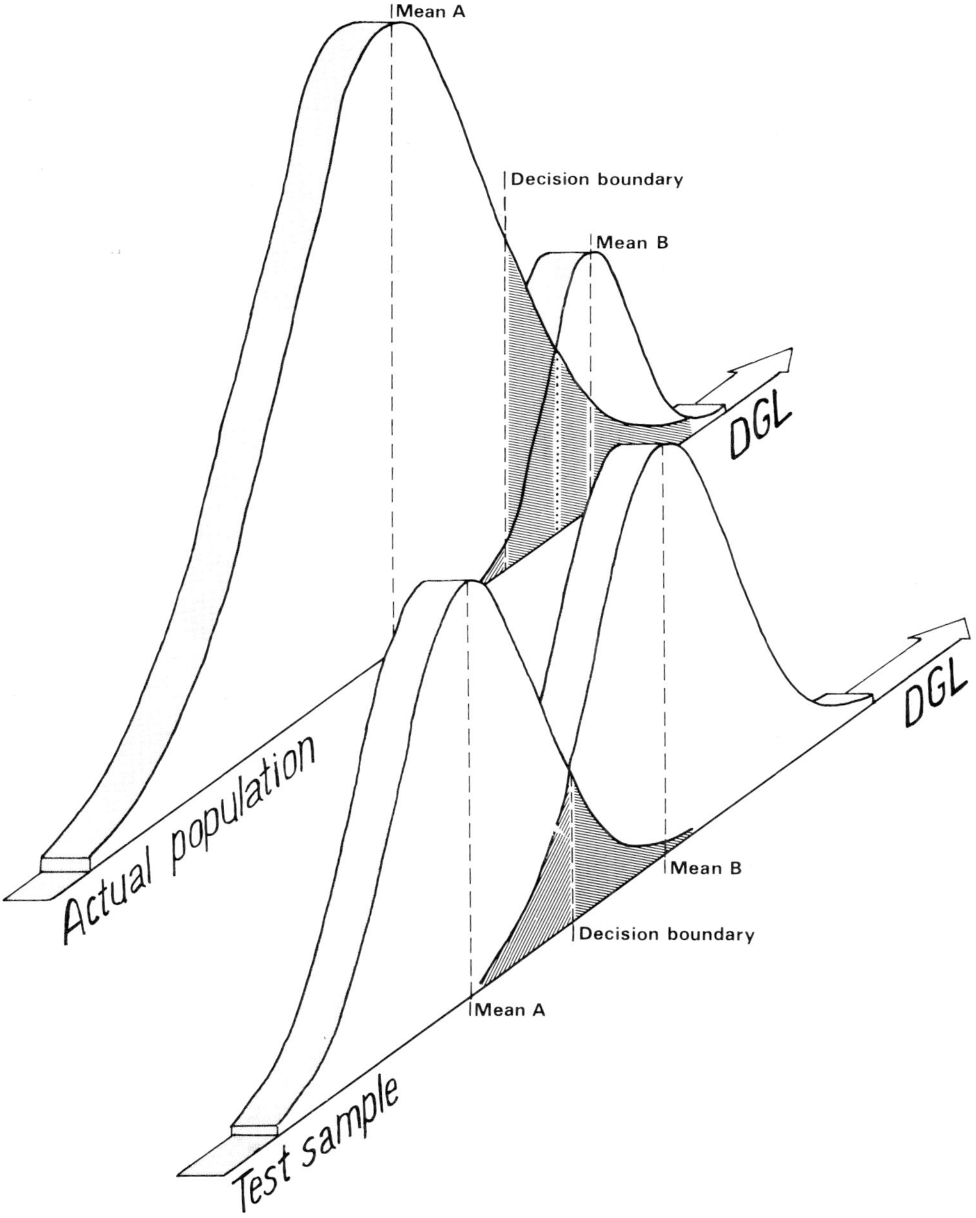

Figure 4.2 Overlap between training samples and actual population. For the former the Y-axis represents frequency in %, and for the latter it represents frequency in absolute number.

overlap between classes, Figure 4.2, decreases when the Euclidean distance between the means increases, and it also decreases when the standard deviation decreases. These effects can be combined to a normalized distance, D_{norm}, defined as (Swain & Davis 1978):

$$\frac{|\bar{x}_1 - \bar{x}_2|}{s_1 + s_2}$$

where x_1 and x_2 are the class means, and s_1 and s_2 are the standard deviations.

A prerequisite when calculating village separation from other classes is that the distribution of the class DGLs is Gaussian. The first step is to find the class with the least normalized distance from village, i.e. with the largest overlap (hatched in Fig 4.2). Respective class on the right-hand and left-hand tails of the village distribution curve, is utilized for calculation of separability.

The area of the overlap, O, is looked up for the D_{norm} values in a standard normal cumulative probability table. And the separability percentage equals

$$100 \times (1 - (O_{left} + O_{right}))$$

With the normalized distance the decision boundary between the classes may also be calculated,

$$\bar{x}_1 + s_1 \times D_{norm} = \bar{x}_2 - s_2 \times D_{norm}$$

The remote sensing literature depicts a variety of more sophisticated distance measures. The Bhattacharyya distance is suggested by Haralick & Fu (1983), and a divergence measure is discussed by Swain & Davis (1978) and Singh (1984). Fu (1976) has a historical review of the development of distance measures. For simplicity reasons (one dimension), the less elegant distance measure, the normalized distance, will be utilized.

As will be noticed in the Section 4.4, a separability measure, of whatever kind, cannot be used to choose the best classification method without precise areal knowledge of the land cover classes. However, it may be used for the culling of additional features to a number of already selected features (see Chapters 5 and 6).

4.3 Classification accuracy estimation

Several different types of measures on the accuracy of classifications has been used for the last decade. Mainly the methods are founded on number of correctly classified pixels with or without involving the pixels in the training areas. More seldom are the objects that are to be classified included in the accuracy calculation as objects, but as a number of pixels.

Much discussion has also been devoted to how many (if any) of the training pixels may be used for accuracy estimation, and how these pixels are to be included in the estimation.

The most important issue is, however, that the analyst knows what classes or what objects are most important to his classification. To use the standard accuracy estimation methods without reserve, without a priori knowledge of class sizes or number of object may give devastating results. Just regard a typical example given by Prof. Stein W. Bie (pers. comm.). Which one of the classified maps in Fig 4.3 are most correct?

Algorithms

Some of the classical algorithms for pixel accuracy estimation might be relevant also for object accuracy

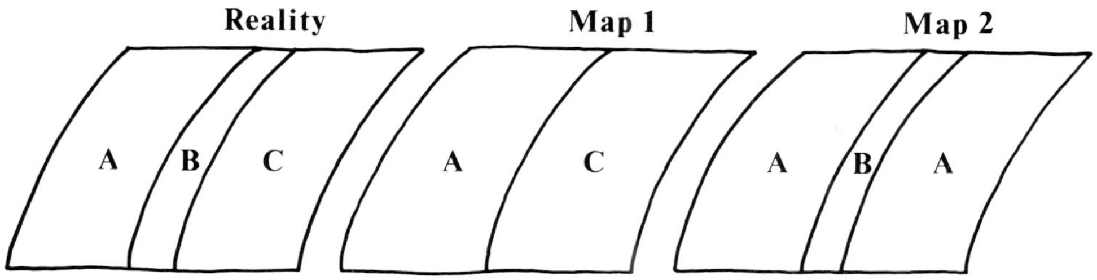

Figure 4.3 Problems of accuracy estimation. A, B and C illustrate 3 different classes. The accuracy depends on the importance of the classes.

estimation.

The most common way to express the mapping accuracy is a simple confusion matrix, which defines how many ground truth pixels from each class that have been classified into each of the classes. The complete matrix, i.e. all classes mixed, or a matrix comprising only one class, might simply be illuminated by a Venn diagram (Fig 4.4), embracing the two subsets MAP and REALITY.

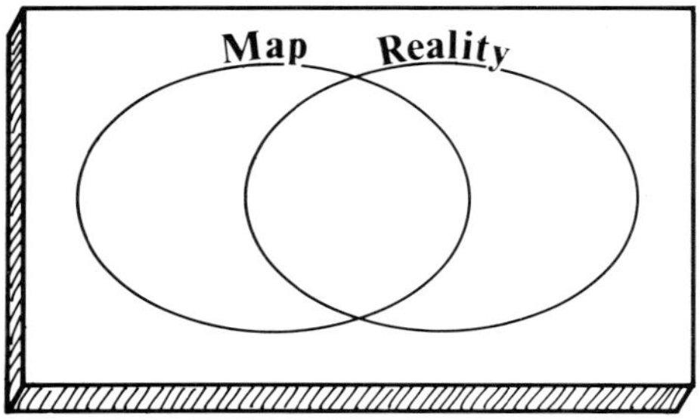

Figure 4.4 Venn diagram depicting the two subsets MAP and REALITY.

Helldén (1980) has a comprehensive compilation of some 'classical' accuracy estimate methods, their advantages and their drawbacks. In Table IV.1 are the formulas described in Boolean terms to fit the set theory of Fig 4.4.

(1) is the exclusively most used formula for accuracy estimation, although it does not take into account the size of REALITY. (1) will be utilized here when the only goal is to include all villages, and over-classification is less important. 'Number of villages' is based on objects while over-classification is based on pixels, thus, they cannot be combined in one accuracy value. Hence, also the the total number of classified pixels is betoken.

When classification procedures are to be compared, formula (4), referred to as 'AND-OR value', is used, sometimes enlightened by Venn diagrams.

Table IV.1 Some 'classical' accuracy estimation algorithms with some of the literature referred to by Helldén (1980).

Algorithm		Reference
(1)	(MAP.AND.REALITY) / MAP	van Genderen & Lock (1977)
(2)	(MAP.AND.REALITY) / REALITY	Helldén (1978)
(3)	(MAP - REALITY) / REALITY	Helldén (1977)
(4)	(MAP.AND.REALITY) / (MAP.OR.REALITY)	Stern (1979)
(5)	2 x (MAP.AND.REALITY) / MAP + REALITY	Hellden (1980)
(6)	geometrical mean of (1) and (2)	Lunden (1977)
(7)	arithmetical mean of (1) and (2)	

4.4 Accuracy prediction

The classification accuracy may be predicted with divergence or J-M distance (Swain & Davis 1978). Input data for the transformed divergence, suggested by Haack (1983), are the means and covariance matrices from a set of training areas. Similar input data is used for all other methods to predict classification accuracy through distance measures.

However, it is important to note that all predictions of classification must involve an a priori knowledge about the cover percentage of all the input classes. Since these percentages are rarely known, accuracy prediction is unfeasible, which the following example enlightens.

Suppose that the village pixels in the study area adopt DGLs in the range 45 to 55 (mean 48.0 and standard deviation 2.0) and that dense Marakh vegetation on reddish sandy soils may be in the range 47 to 57 (mean 52.0 and standard deviation 2.0) for the same band.

Using the normalized distance, the predicted classification accuracy is 84.1 % for both classes, since the algorithms do not consider the sizes of the classes. But since the area of the Marakh vegetation is likely ten times larger than that of the villages, the actual classification accuracy is 34.6 % for the village class and 98.1 % for Marakh.

4.5 Classification algorithms

Remote sensing literature generally contain descriptions of calculation procedures to find decision boundaries and to assort each pixel after its most likely class affiliation. The boundaries are usually extracted from training areas representative for their classes.

The simplest case, just to define the class boundaries in 1-D, is usually called density slicing. When using two or more bands, there are three algorithms that are more commonly used than others, parallelepiped, minimum distance and maximum likelihood (M-L) classification. For definitions see e.g. Schowengerdt (1983).

M-L classification is a much more complicated and expensive computer process than is the parallelepiped classification, and it uses probability functions including a prori knowledge of land cover percentages. Strahler (1980) discusses the importance of this knowledge for the M-L classifier.

The effectiveness of the different classification algorithms is sometimes debated in remote sensing literature, but seldom thoroughly investigated. However, it is most often experienced that the accuracy is more conditioned upon the selection of class boundaries and the representativity of the training areas than on the choise of algorithm.

The recent years have seen several new classification algorithms born, methods speeding up computations, methods incorporating neighbours to enhance the accuracy, etc (e.g. Saebø et al 1985). Since programmes for such fancy methods were not available they were not tested in the present investigation.

4.6 Classification philosophy

The objectives of this chapter is to reveal

- the number of spectral bands needed for classification;

- the best band combination for that number of bands;

Remembering that there are 254 ways to select band combinations among the available eight MSS bands, the study was divided into two different approaches

1) Preliminary assorting, using 1 to

4 bands to pick the best band combination for spectral and bi-temporal classification using a parallellepiped classifier.

2) Finer assorting using 1 - 8 bands to pick the number of bands needed to fulfil the needs. Here the inferences of the preliminary classifications are used to decrease the computational efforts using the M-L classifier.

It is important to note that for all classifications only one class was used, namely villages. Otherwise, the 'probability' of a spectrally juxtaposed class, which is likely much larger than the village class, would relentlessly intrude upon the village class 'probability' (Cf Fig 4.2).

The use of only one class will neglect the benefits of M-L classification, and instead suggest a minimum distance classifier where the Euclidian distance, weighted by the standard deviation, limits the class labelling. However, such a classifier was not available in Lund where this part of the study was performed, but the M-L classifier will have the same effect, though with a considerable waste of computing time.

The preliminary assorting was made with a simulated parallellepiped classifier, simply by logically

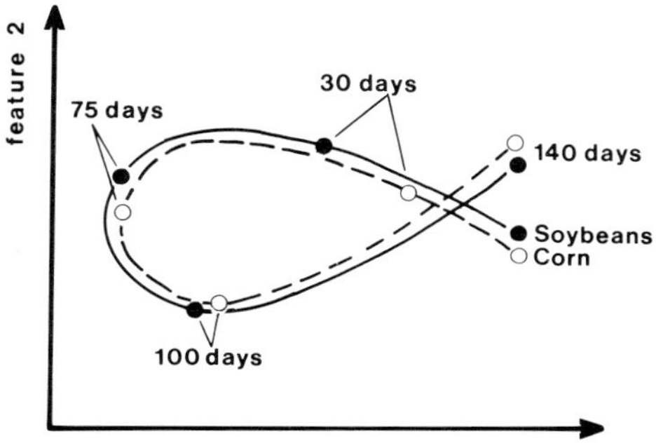

Figure 4.5 Temporal change of vegetation reflectance (after Landgrebe 1971).

OR-ing density slices from each of the 8 bands, an approach which is tasteful considering that 154 band combinations were tested.

The presented approach will, thus, imply a conscious over-classification to be sure to embody all the villages, and the surplus will hopefully be reduced in the object classification described in Chapter 7.

4.7 Multi-temporal approach

Multi-temporal data indicates that it emanates from more than one moment of acquisition. Figure 4.5, which is borrowed from Landgrebe (1971), shows how the spectral response of two vegetation types changes by time.

Temporal changes of reflectivity may be natural, as when the leaves turn in the autumn, or dependent on human activities, such as plowing or harvesting. The changes may be short-lived, such as soil moisture or long-lived as erection of a new village, and it may be caused by a cyclic diurnal phenomenon, such as sun-angle, or by a multi-annual one, as e.g. fallow period.

Table IV.1 displays the seasonal cycle regarding different phenomena in Kordofan and the dates of the available Landsat data.

When using time as a discriminant, it is interesting to apprehend the optimal dates for data aquisition. However, these dates are rarely known, and even if they were, acquiring satellite data from these dates may be hard.

The interval between the dates should be prolonged enough to enable a change of the appearance of the study object, but short enough to unable an alteration of the state of the object. For example, the village surroundings will change its reflectivity at the turn of the season, which makes it appropriate to use

Table IV.1 Season cycle in Kordofan (from Reeves & Frankenberger 1981).

Seasonal Variables		Mar	Apr	May	June	July	Aug	Sept	Oct	Nov	Dec	Jan	Feb
1. Season		Şeef (hot, dry)		Rushaash (early rains)		Khariif (rain)		Darat (harvest)		Shita (cool, dry)			
2. Temperature C°	Minimum	17.1	20.5	23.2	23.6	22.5	21.6	21.4	21.1	17.0	13.4	18.1	13.3
	Maximum	39.4	41.3	42.2	40.2	36.8	38.8	36.9	38.6	36.6	35.0	35.2	36.1
3. Rainfall mm		1.5	1.5	22.4	19.4	100.7	133.9	49.9	16.3	1.5	0	0	0
4. Wind Patterns		Northerly Winds		High Winds (Habuub)		Southerly Monsoon				Northerly Winds			
5. Access to Water		Difficult Expensive				Readily Available From Rain Pools				Available From Hafiirs, Shallow Wells, Govt and Private Deep Wells			
6. Food Consumption		More Sorghum Eaten Than Millet, Less Meat				More Millet Milk & Milk Products, Vegetable, From Farm & Market		Mariisa (beer) & Meat, Fariik (immature millet & sorghum)		Millet & Meat Mariisa (sorghum beer)			
7. Gum Production		Collecting Gum From Young Trees, Sometimes From Old Ones							Tapping Old Trees		Tapping Young Trees & Collecting Gum From Old Trees		
8. Cropping Cycle	Millet		1st Planting	2nd Planting					Cutting			Land Clearing	
		Land Clearing			1st Weeding		2nd Weeding			Threshing			
	Sorghum			1st Planting		2nd Planting			Cutting			Land Clearing	
		Land Clearing			1st Weeding		2nd Weeding			Threshing			
	Sesame			1st Planting	2nd Planting				Cutting			Land Clearing	
		Land Clearing			1st Weeding		2nd Weeding			Threshing			
	Groundnuts				1st Planting				Pulling			Land Clearing	
		Land Clearing			1st Weeding		2nd Weeding			Threshing			
9. Farm Labor		Land Clearing		Planting Before Rains (rimeel) After Rains (nadaaya)	← 1st Weeding	Weeding	2nd Weeding →		← Cutting and Threshing →			Land Clearing Begins →	
10. Crop Marketing		Farmers Sell off all Remaining Cash Crops (a) To Buy Feterita (b) To Buy Animals (c) To Pay for Land Cleaning, Planting & Weeding				Crop Marketing Generally Dormant; Local Sale of Seed for Planting			Sale of Early Maturing Crops to Merchants to Pay for Commodities at Shops	Continued Selling of Crop at Shops to Buy Consumption Goods; Increasing Sales of Cash Crops in Large Quantities at Rural Markets			
11. Livestock Marketing		Sale of Weak and Dying Animals, Particularly Goats				Period of the Greatist Number of Sales, for all Livestock, by Farmers and Nomads. Farmers Use Earnings from Sales to Finance Hiring of Farm Labor and to Pay for Household Expenses, Nomads Sell off Weak Animals and Buy Better Quality Livestock					Horse Sales (for hauling water and crops); Sale of Other Livestock Generally Dormant		
12. Transportation		Loose Shifting Sand Makes Travel Difficult				Roads Often Impassable Due to Rains					Crop Trucks Travel Frequently from Villages to Urban Markets		
13. Off Farm Income Sources		Migration for Wage Labor Off-Farm Occupation in Village									Migration for Wage-Labor, Off-Farm Occupation in the Village		
14. House Construction		Building and Maintenance of Houses and Compounds									Building and Maintenance of Houses and Compounds		

multi-seasonal images. However, it is inappropriate to use images acquired several years apart, since the village may have changed in size considerably during that period of time. Such an image combination is, on the contrary, an instrument to study the change of the state of known villages.

Theoretically, very small differences should be found between village response from the two dates of available data. The only seasonly changes inside the village are the shade trees, which are few and change moderately. The differences are caused by atmospheric effects, and maybe also the radiometric and geometric corrections made at the data pre-processing at EROS Data Center.

When absolute reflectance of objects are needed or when knowledge of grey tone levels is transferred from one scene to another radiometric calibrations is needed. For the present study no radiometric calibrations were needed.

Landsat data from the wet season and the dry season 1979 over the study area allowed simultaneous tests of multi-spectral and multi-temporal data, or rather bi-temporal data. The bands were geometrically registered with a nearest neighbour resampler on the ARIES II system.

Regarding the result of Chapter 3, one would suggest that a cubic convolution resampling is to be preferred rather than nearest neighbour resampling. However, it would also imply a greater difference between the two images. The cubic convultion will be used later when village areas are more precisely determined.

4.8 Spectral clusters

The spectral response of villages is already exhibited in Chapter 3, and other conceivable classes are functions of chiefly two parameters, soil colour and vegetation cover.

From Table IV.2 it is easily comprehended that most classes will be

Table IV.2 Mean DGLs (and standard deviation) from a sample of training areas (200 - 2000 pixels in size) from some of the classes in the investigation area.

Class	Mar4	Mar5	Mar6	Mar7
Village	28.1 (1.4)	47.9 (3.0)	55.6 (3.2)	48.1 (2.9)
Qoz (bare sand)	37.3 (1.5)	73.5 (3.5)	82.8 (4.3)	69.1 (3.3)
Qoz (dark veg.)	29.8 (1.2)	46.3 (1.9)	51.8 (3.0)	45.4 (2.4)
Tin (with bush)	37.0 (1.9)	58.0 (3.6)	62.2 (3.8)	54.0 (3.6)
Gardud	24.6 (1.9)	32.2 (2.0)	32.8 (2.5)	25.4 (2.6)
Class	**Sep4**	**Sep5**	**Sep6**	**Sep7**
Village	27.9 (1.4)	43.5 (3.3)	55.4 (2.9)	45.1 (2.9)
Qoz (bare sand)	42.0 (3.1)	79.3 (8.5)	95.4 (5.1)	73.4 (2.7)
Qoz (dark veg.)	28.6 (1.2)	42.2 (2.1)	53.4 (2.9)	44.8 (2.5)
Tin (with bush)	35.7 (3.1)	50.2 (6.4)	62.3 (4.2)	51.3 (2.8)
Gardud	23.1 (1.3)	27.4 (1.9)	30.4 (2.1)	22.4 (2.0)

readily excluded, whereas it also may be suspected that others cannot be separated solely by a single-band discrimination.

Henceforth are the March MSS bands called Mar4 etc, and the September bands are referred to as Sep4 etc.

Table IV.3 Result of the separability test (normalized distance) of village against the training areas in Table IV.2.

Mar4	59.8 %	Sep4	64.7 %
Mar5	56.6 %	Sep5	10.5 %
Mar6	55.6 %	Sep6	46.7 %
Mar7	51.4 %	Sep7	38.6 %

After a rough look at the image, the classes of Table IV.2 could certainly be regarded as all existing classes. The training areas may also be regarded as representative for these classes. Hence, if one believes in accuracy prediction, the percentages in Table IV.3 equal classification accuracy. However, the separability vacillated on small changes of the training areas, sometimes more than 20 % using different samples from the same training area. This, again, calls the attention to the importance of representative samples.

4.9 Preliminary feature selection

For each of the eight bands the mean and the standard deviation for all village pixels in the study area were calculated. All bands were sliced according to village mean DGL +- one standard deviation. These eight slices were stored as binary images, and could then be logically OR-ed in any combination, which equals parallellepiped classification with the selected combination of bands.

One standard deviation on either side of the mean statistically entails that 68 % of the village pixels are classified as villages, when using one band. Using more bands, some more village pixels may be lost, but a plausible assumption is that roughly the same number of villages pixels are lost for the same number of bands. Thus, may the best band combination be selected on the basis of the least over-classification.

Single-band classification

Originally, the village pixels comprised only about 0.7 % of the test area. All other classified pixels are by necessity misclassified. Table IV.4 indicates the overestimation for each band.

Table IV.4 Pixels classified (in % of total area) by a single-band parallelepiped classifier. These values may be compared with the values in Tables IV.5-IV.7.

Band	% of total area
Sep6	10.43 %
Sep7	12.30 %
Sep5	16.25 %
Mar5	20.65 %
Mar6	23.38 %
Mar7	24.28 %
Mar4	25.05 %
Sep4	34.77 %

The inference to draw from IV.4 is that villages are immensely difficult to discriminate by single-band classification because of their minute area relative to other classes.

Two other points ought to be discussed while having these tables in memory.

Colvoresses (1977) states that MSS 5 is the fundamental band for indicating boundaries between natural and cultural classes and that MSS 6 and 7 have demonstrated operational value for cultural delineation.

According to Table IV.2 and IV.4, it seems that MSS 6 is slightly more beneficial than MSS 5 and 7, but extensively better than MSS 4 for single-band classification. If this is to be regarded as a petty discrepancy to Colvoresses, it might be explained by the fact that these Sudanese villages are composed of natural materials in lieu of concrete as larger cities usually are.

In contrast to these results, Reining (1974) found that dry season imagery usually, but not always, is better than wet season imagery for settlement interpretation in her Sahelian test area. One explanation is found in the disparity between the requirements for manual interpretation and automatic discrimination of villages, which is further discussed in Chapter 9.

Multi-band classification

Only 4 bands were used simultaneously in the preliminary feature selection. The theory that more bands are unnecessary, which is suggested by Fig 4.1, will be tested later. An additional motive for restricting the test to 4 bands is the limiting of manual/computer exertion when combining the slices with the Boolean operator.

The multi-band test comprised three levels of parallellepiped classification, two-band, three-band and four-band classifications. The same decision boundaries as for the single-band classification was used.

Table IV.5 Pixels classified (in % of total area) by a two-band parallelepiped classifier, for single-season and multi-season combinations. Only the best combinations shown.

Dry season		Wet season		Multi-season	
Mar6 Mar4	8.73 %	Sep7 Sep5	4.95 %	Sep6 Mar4	4.51 %
Mar7 Mar4	8.76 %	Sep6 Sep5	5.55 %	Sep7 Mar4	4.77 %
Mar5 Mar4	9.17 %	Sep7 Sep6	6.64 %	Sep6 Mar5	5.94 %
Mar7 Mar5	15.53 %	Sep7 Sep4	7.46 %	Sep6 Mar7	6.04 %

There are $\binom{n}{k}$ possibilities to select n features out of k features, which entails 28 two-band, 56 three-band and 70 four-band combinations when having a total of 8 bands.

The entire study area was used for all the 154 classifications, and Table IV.5 indicates the number of pixels classified as a percentage of the total number of pixels in the investigation area.

Before any conclusions are extracted from Table IV.5, also Table IV.6 and IV.7 ought to be considered.

Studying the next tables it must be remembered that only 4 single-season three-band combinations $\binom{3}{4}$ from each season exist, whilst 48 bi-season combinations $\binom{3}{8}$ can be constructed, and consequently only one single-season four-band combination $\binom{4}{4}$, but 68 bi-season combinations $\binom{4}{8}$ are possible.

Table IV.6 Pixels classified (in % of total area) by a three-band parallelepiped classifier, for single-season and bi-season combinations. Only the best combinations shown.

Dry season	
Mar7 Mar5 Mar4	6.85 %
Mar6 Mar5 Mar4	7.22 %
Mar7 Mar6 Mar4	7.41 %
Mar7 Mar6 Mar5	14.48 %
Wet season	
Sep7 Sep6 Sep5	3.53 %
Sep7 Sep5 Sep4	4.43 %
Sep7 Sep6 Sep4	4.80 %
Sep6 Sep5 Sep4	5.00 %
Multi-season	
Sep7 Sep5 Mar4	2.10 %
Sep6 Mar7 Mar4	2.20 %
Sep7 Mar7 Mar4	2.27 %
Sep6 Mar6 Mar4	2.37 %
Sep7 Mar6 Mar4	2.38 %
Sep6 Sep5 Mar4	2.39 %

Table IV.7 Pixels classified (in % of total area) by a four-band parallelepiped classifier, for single-season and multi-season combinations. Only the best combinations shown.

Dry season	
Mar7 Mar6 Mar5 Mar4	6.31 %
Wet season	
Sep7 Sep6 Sep5 Sep4	3.18 %
Multi-season	
Sep7 Sep5 Mar7 Mar4	1.04 %
Sep7 Sep5 Mar6 Mar4	1.18 %
Sep6 Sep5 Mar7 Mar4	1.27 %
Sep6 Sep5 Mar6 Mar4	1.35 %
Sep7 Sep6 Mar7 Mar4	1.36 %
Sep7 Sep5 Mar5 Mar4	1.37 %

The first conclusion to make is that multi-season imagery is more powerful than is the single-season, which was expected and agrees with Reining (1974).

The most striking result is that the March MSS4 band, which is rather incapable of discriminating villages per se, becomes invaluable in combination with the longer wave-length bands of September. This is to be explained by the fact that MSS 4 confuses large amounts of bushy vegetation on silty soils with villages, while the September MSS 6 and 7 regard Marakh vegetation on sandy soils as villages, but not vice versa. This coincides with e.g. Haack (1983) who states that

a) the best one-band or two-band combinations are not always subsets of the best three- and four-band combinations;

b) the best combinations generally include bands from different major regions of the electromagnetic spectrum.

Table IV.8 Bands combinations used for classification tests.

1 band	- Sep6
2 bands	- Sep6, Mar4
3 bands	- Sep7, Sep5, Mar4
4 bands	- Sep7, Sep5, Mar7, Mar4
5 bands	- Sep7, Sep5, Mar7, Mar5, Mar4
6 bands	- Sep7, Sep6, Sep5, Mar7, Mar5, Mar4
7 bands	- Sep7, Sep6, Sep5, Mar7, Mar6, Mar5, Mar4
8 bands	- all

4.10 Feature reduction test

Based on the conclusion from the preliminary feature selection the preferable band combinations were passed on to the feature reduction test, where the discrimination power of each combination was studied. The test sub-area used included 22 villages.

Table IV.8 denotes the bands combinations used for the M-L classification. 1- to 4-band combinations are selected directly from the preliminary selection. The 8-band combination is obvious, and for 5 to 7 bands the general conclusions of the test above is used.

The first M-L classifications revealed that the used mean + 1 standard deviation meant a rapid decrease in number of correctly classified objects when increasing the number of features. It also indicated that the first villages to disappear were those with a lighter appearance. Hence, three additional M-L classification tests were performed, one with mean + 3 standard deviations and two similar classifications with a mean adapted to lighter villages only. The result of the tests can be seen in Figure 4.6.

4.11 Conclusions and discussion

The major deduction to draw from the preliminary feature selection was the advantage of choosing features from different wave-length bands and from different seasons of the year. The band combinations used in the succeeding feature reduction test are denoted in Table IV.8.

Striking at first glance is the fast decline of the number of correctly classified villages when increasing the number of bands. However, it coincides with Fig 4.1b.

For a Gaussian distributed 1-D village pixel population only 68.2 % is located within one standard deviation. For two dimensions 46.5 % of the population lies within the one standard deviation limits, while for three standard deviations the same figures are 99.8 and 99.6, respectively. This reasoning can easily be extrapolated to eight dimensions.

The loss of a certain percentage village pixels does not imply a corresponding loss of villages as objects, which is explained by the fact that the centre pixels (which are close to the mean) will remain classified and the lighter fringe mixels are lost. However, when the percentage becomes too small also entire villages are lost, which thus is the major cause of the accuracy when using more bands.

It must also be remembered that the over-classification declines much more rapidly than the villages, because of the measurement in number of pixels in this test.

A correctly executed M-L classification involving all existing classes would not suffer from these problems and the overall classification accuracy would probably not

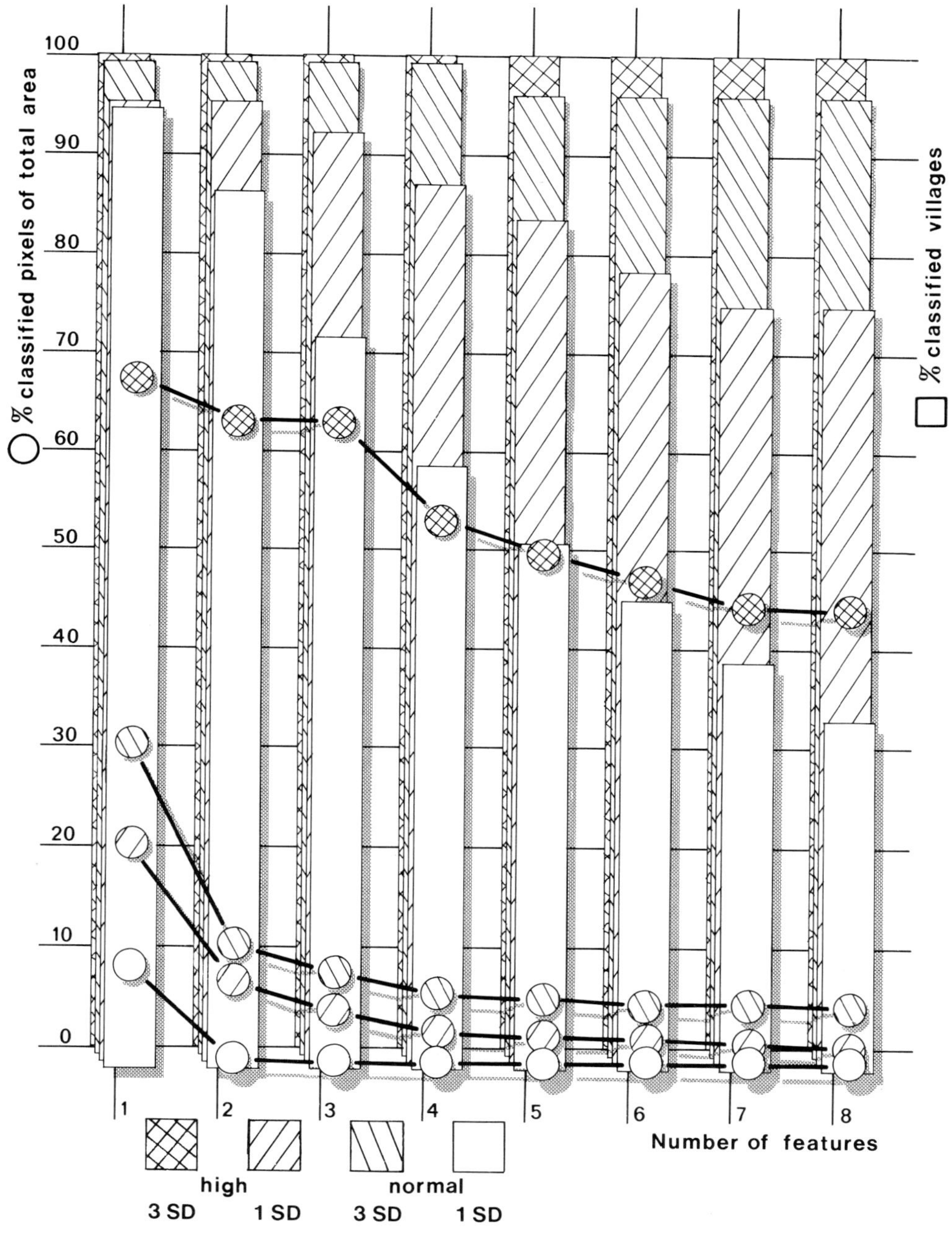

Figure 4.6 Result of the final classification with 1 to 8 bands. Circles represent pixels classified as village in percent of total area. Bars represent correctly classified villages.

decrease for an increased number of bands. However, for reasons mentioned earlier, the village classification accuracy would not profit from the M-L approach.

The goals to keep all villages classified and to keep a low over-classification can be achieved only by using normal mean + 3 standard deviations, and 4 bands (Fig 4.6). Thus, the most powerful 4-band combination will proceed to the combined automatic tests.

The problem of over-classification was specially studied on a field work, where the areas that were steadily spectrally confused with villages were visited. The major part of them turned out to be young dense stands of acacia senegal on silty soils.

The most dubious areas were picked out as training areas for separability tests when selecting arithmetical features (Chapter 5) and texture features (Chapter 6), which were chosen only depending on their ability to separate these acacia stands from villages.

The use of principal components was noy tested here, but will untested be incorporated in the combined automatic tests (Chapter 8).

5 arithmetical indices

Arithmetical functions, whereof the exclusively most common one is ratio, of different MSS bands are used mainly for three purposes:

- to monitor green biomass;

- to suppress topographic effects or atmospherics;

- to discriminate between spectrally similar classes.

Practically, it is a simple operation, e.g. one band is divided by another, pixel by pixel. To remain in the range 0-255 for the output pixel values, they are usually normalized in one way or another.

The output features may then be input features to e.g. a maximum likelihood classification, or combined by other features to a false colour composite for manual interpretation.

5.1 Related literature

In the remote sensing literature, logical explanations are rarely found to why a given function is used, or why given bands are put into the function. The functions are not arbitrarily chosen, but rather empirically. And scarcely nowhere it is theoretically described how the human eye or brain would perceive the effect of a given function.

This neglect is startling since a lot of papers report successful ratios, but understandable since it is an area with immense complexity involved.

Abrams (1980) defined ratioing as the slope of the spectral curve between two bands, but also mentioned the disadvantage that ratios might confuse dissimilar classes having similar spectral slopes but different albedos.

Swain & Davies (1978) pointed out that ratio indices may not be Gaussian and hence rejected for M-L classification.

Biomass indices

Supported by the radiation theory, MSS 7 / MSS 5 and MSS 7 - MSS 5 are ordinarily employed for vegetation studies. Green vegetation has a maximum absorption of light at the wave-length 0.45-0.65 microns (corresponding to MSS 5) and a maximum reflection at 0.7-1.3 microns (corresponding to MSS 7).

Wiegand et al (1974) suggested MSS 5 / MSS 7 or MSS 7 - MSS 5, displayed with the 'pure' MSS 6 and MSS 7 bands, as practical for manual studies of vegetation cover and density.

Rouse et al (1973) proposed

MSS 7 - MSS 5 / MSS 7 + MSS 5

as a vegetation index.

According to Tucker (1979), the linear greenness index

$$-0.29 \times MSS\ 4 - 0.56 \times MSS\ 5 + 0.60 \times MSS\ 6 + 0.49 \times MSS\ 7$$

was proposed by Kauth & Thomas (1976).

Sabins (1978) mentions the ratio

(MSS 4 - MSS 5) / (MSS 4 + MSS 5)

which also has been used by, e.g. Olsson K. (1985) for estimation of woody biomass, which is different from green biomass.

Siegal & Goetz (1977) found that MSS band ratios 4/6, 4/7, 5/6 and 5/7 are significantly decreased with increased amount of vegetation. Ratios 4/5 and 6/7 were less influenced and are thus more diagnostic of the soil.

Topographic and atmosperic effects

Taranik (1978) recommended 4 / 5, 5 / 6 or 6 / 7 to correct atmospheric effects. He, as well as Sabins (1978), gave a rationale for illumination differences equalized by ratioing.

Lindell (1983), inter alia, suggested the chromaticity indices

MSS 4 / MSS 4 + MSS 5 + MSS 6 and
MSS 5 / MSS 4 + MSS 5 + MSS 6

which describe the colour of the object, but simultaneously attempts to eliminate atmospheric disturbances and the need for radiometrical calibration of Landsat data.

Townshend (1981) noticed that addition and multiplication are infrequently used, but multiplication, when tried, tends to enhance the ruggedness of the terrain.

Spectral discrimination

Taranik (1978) suggested temporal ratioing, in this case the ratio of two MSS 5 images acquired two years apart, to study changes of strip mine development in Idaho.

Townshend (1981) asserted that ratios enhance subtle contrasts difficult to detect in the original images, especially when bands from different parts of the spectrum are utilized. Referring to Holben & Justice (1980), he also stated that ratios reduce effects of unequal slope illumination, but banding is enhanced.

5.2 Index selection

Mathematically, the functions are easily comprehended; subtraction and ratioing, etc, are simple operations. The effect of a function applied on an image, is also easy to illustrate (Fig 5.1). But when it comes to selecting one function among several, the difficulties

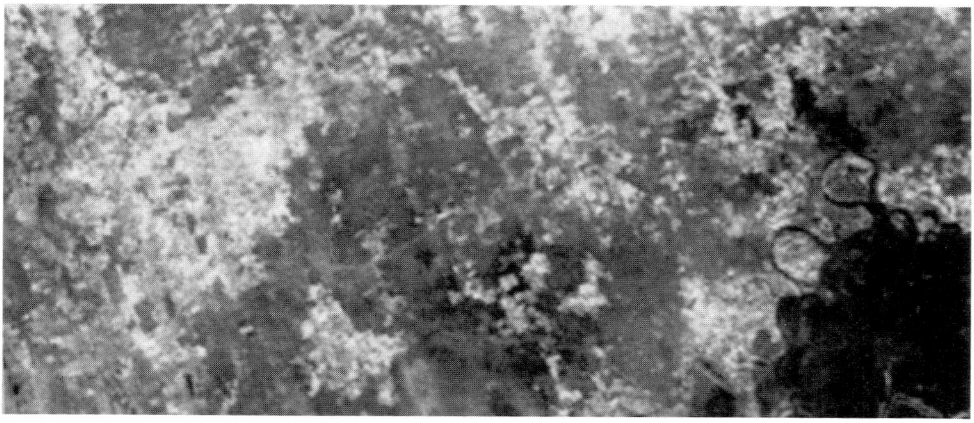

Figure 5.1 a) Sep5 of the sub-area in Figs 5.1 c - e.

Figure 5.1 b) Sep7 of the sub-area in Figs 5.1 c - e.

Figure 5.1 c) Sep5 x Sep7 of the same area. Note the close resemblance to the original bands.

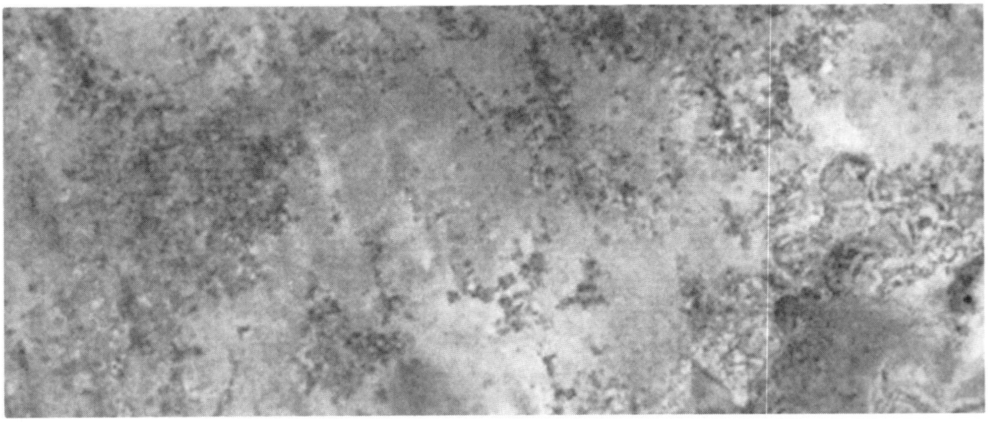

Figure 5.1 d) Sep7 / Sep5 of the sub-area in Figs 5.1 a) and b).

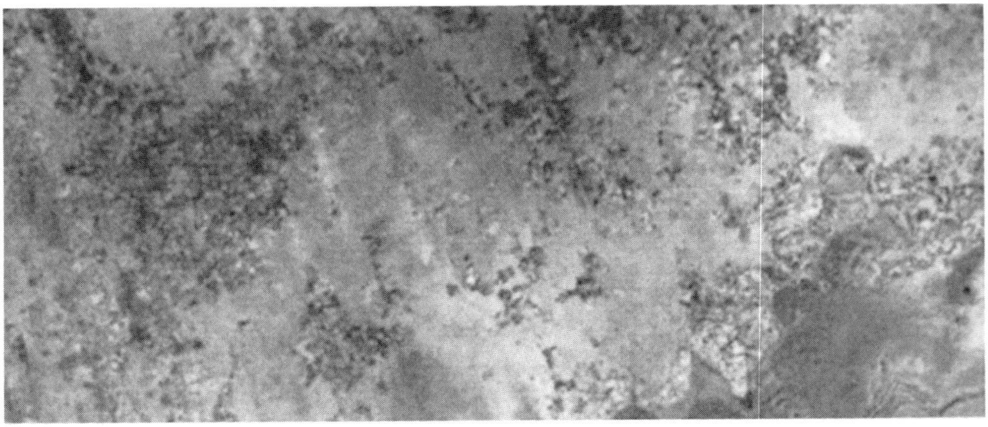

Figure 5.1 e) Sep7 - Sep5 of the same area. Note the petty differences from Fig 5.1 d). This similarity may also be seen in Figs 5.3 b) and c).

start.

As indicated above, the most common methods of choosing arithmetical indices are based on the theories of physics, but the very selection is done empirically.

For example, the high reflectances of vegetation in MSS7 and the low reflectances in MSS5 are well established in the laws of physics. But does the discrimination of green, fresh vegetation from less vegetated areas demand a subtractive index, or is a ratio more powerful? The physical laws do not assist in that selection, and consequently, all kinds of indices that enhance the differences between MSS7 and MSS5 are used (Cf the literature review above).

Olsson K. (1985) compared a variety

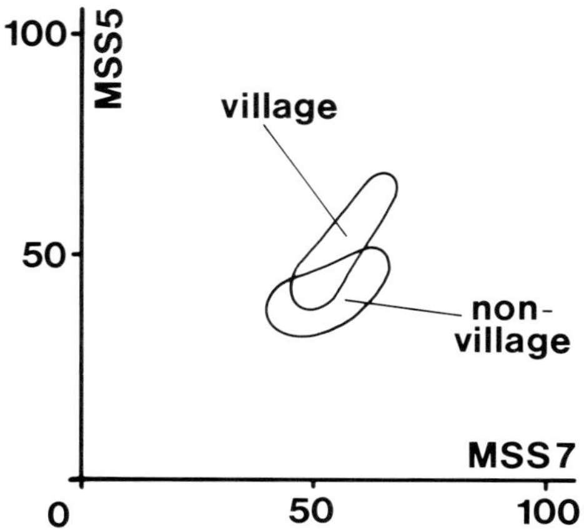

Figure 5.2 Typical scattergram for training areas of the classes used in the separability test.

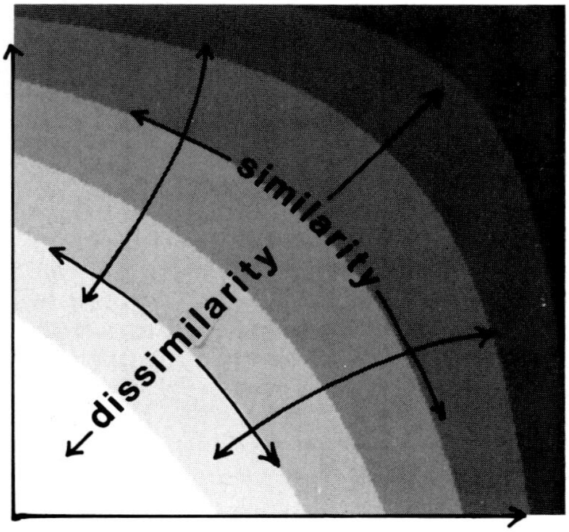

Figure 5.3 a) Graphical model of MSS7 x MSS5.

of vegetation indices, and she found that the indices have an extremely high correlation, implying that it makes no difference which index to use. This is likely to be true for vegetation, but not necessarily true for all kinds of index applications.

If all the classes were linearly distributed in the spectral space, the maximal separation could be achieved through a canonical trans-

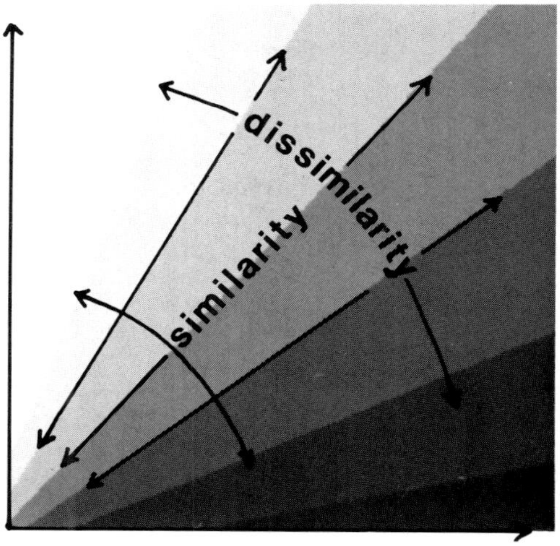

Figure 5.3 b) Graphical model of MSS7 / MSS5. Probably the best function for separating the two training areas of Fig 5.2.

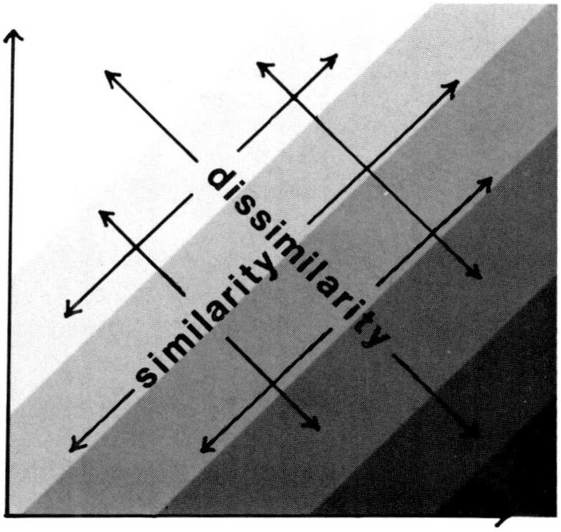

Figure 5.3 c) Graphical model of MSS7 - MSS5.

formation (see Chapter 4), and no arithmetical indices were needed. However, such an assumption cannot be accepted.

Graphical approach

A prefatory approach may be to compare a scattergram comprising the training areas with graphical models of some basic indices. Fig 5.2

denotes a scattergram with some training areas, and Fig 5.3 depicts some graphical models.

Just as spectral scattergrams are insufficient for selecting spectral features, the graphical models are insufficient for choosing arithmetical features, but they are appropriate for an initial understanding of the problem. When utilizing more than two dimensions, the graphical approach becomes unfeasible.

The 'similarity' arrows in the models (Fig 5.3) describe the preferable spectral orientation of the training areas in the scattergram, if that index is to be used. The 'dissimilarity' arrows describe the preferable direction of separation of the training areas.

Although the indices are difficult to understand, it is possible to calculate the separability, not only for different functions, but also for different input data to the functions.

5.3 Separability

For a mathematical background to the separability test, read Section 4.2.

Indices

This study unprejudicely tested 880 different arithmetical indices, with four features (or less) as input data. Certainly, more input bands would be interesting, but would also boundlessly increase the number of indices.

There are, of course, many more than 880 possible 4-dimensional arithmetical indices, the ones used consist of 22 sub-functions in the numerator and 40 in the denominator (22 x 40 = 880).

The 22 numerator sub-functions comprise six two-band subtractive ones, all 12 possible two-, three- and four-band multiplicative ones, and the four single-band sub-functions.

The denominator was used for the four single-band and some two-, three- and four-band subtractive and additive sub-functions. Of the latter type, 64 possibilities exist, but they were reduced to 36, since e.g. MSS 4 - MSS 5 and MSS 5 - MSS 4 offer the same separability. The indices were chosen in a manner that the same sub-functions did not appear above and below the line at the same time.

These 880 functions were tested on 20 different 4-band combinations, which totalled more than 5,000 combinations. Firstly, the single-season band combinations were tested, secondly the best combinations of the multi-temporal classification, and thirdly, an arbitrarily chosen sample of 70 possible combinations (Table V.1).

Training areas

The arithmetical indices were tested for their power to separate villages from spectrally juxtaposed classes, mainly young stands of acacia senegal (see Section 4.11). This approach was based on the concept that arithmetical features should only be utilized in addition to the spectral features, i.e. they are tested solely for problems faced at the spectral classification.

Among non-villages areas, which no spectral feature combination could sort out, six training areas were chosen, and put into the separability test, together with a representative village sample.

5.4 Conclusions and discussion

On the whole, it is not trivial, mathematically or graphically, to select appropriate indices. Presumably, it would be feasible to develop tasteful methods to describe the index separability differences in a more sophisticated mathematical

Table V.1 Separability power of the tested arithmetical indices. Only the best indices shown, together with some interesting comparisons.

a) Best dry-season non-ratio indices	
Mar4 - Mar6 + Mar7	61.72 %
Mar4 - Mar5 - Mar6 + Mar7	59.80 %
Mar6 - Mar7	58.25 %
Mar4 - Mar5 + Mar7	58.25 %

Mar4 - Mar7	17.39 %
Mar4 - Mar6 - Mar7	14.67 %
Mar4 + Mar5 - Mar7	3.99 %
b) Best dry-season indices	
Mar5 / Mar4 - Mar6 + Mar7	64.73 %
Mar5 / Mar4 + Mar5 - Mar6 + Mar7	64.73 %
Mar5 / Mar4 - Mar5 - Mar6 + Mar7	64.73 %
c) Best wet-season non-ratio indices	
Sep4 - Sep7	25.86 %
Sep4 - Sep5 + Sep6 + Sep7	25.23 %
Sep4 + Sep7	24.75 %

Sep5 - Sep7	10.10 %
Sep4 + Sep5 + Sep6 - Sep7	2.40 %
d) Best wet-season indices	
Sep6 - Sep5 / Sep4 - Sep6 + Sep7	54.31 %
Sep6 - Sep5 / Sep4 - Sep5 + Sep7	54.31 %
Sep6 - Sep4 / Sep5 - Sep6 + Sep7	53.52 %
Sep6 - Sep4 / Sep4 - Sep5 - Sep7	53.52 %
e) Best multi-season indices	
Mar7 - Mar6 / Mar7 - Sep4 + Sep5	62.10 %
Mar7 - Mar6 / Mar6 - Sep4 + Sep5	62.10 %
Mar7 - Mar6 / Mar4 + Sep5	62.10 %
Mar7 - Mar6 / Mar4 + Mar6 - Mar7 + Sep5	62.10 %

manner, but it is far beyond the scope of a report in applied remote sensing.

The following conclusions are valid only for village discrimination in the current investigation area. The validity of the conclusions can most unlikely be regarded as general.

An unforeseen result was that the use of March bands alone seemed more powerful than multi-temporal combinations. However, the difference was small and not significant.

Usually, indices involving ratios were superior to non-ratio indices.

It might be noted that the worst multi-temporal 4-band combination in the spectral classification (not denoted in Table IV.7), Mar6-Mar7-Sep4 -Sep5, became the most beneficial one, when put into an arithmetical index.

The conclusive inference was, however, that a large amount of indices behaved similarly, i.e. the differences of their separabilities were not significant. Hence, many of them could qualify for further tests of village discrimination with the combined automatic methods. Mar5 / (Mar4 + Mar5 - Mar6 + Mar7) was chosen arbitrarily among the best indices. To what extent it is useful as an additonal feature cannot be determined prior to the combined classification (Chapter 8).

6 textural discrimination of villages

6.1 Background

Spatial-spectral relationships

The preceding chapter dealt with different concepts of spectral separation, i.e. considering pixel-by-pixel grey-tone variations. Classifiers employing only this kind of information are sometimes referred to as per-point classifiers. The present chapter will discuss the spatial-spectral relationships between pixels.

Gurney & Townshend (1983) defined two common expressions regarding spatial relationships:

- Context: spatial relationships (Fig 6.1) with pixels in the remainder of the scene;

- Texture: spatial variations of pixel DGLs within a contiguous area.

The concept 'spatial relationships' usually implies relations between objects in an image, but may sometimes regard only the pixels in the image.

Figure 6.1 illustrates the fundamental types of contextual relationships. Connectivity is a property sometimes employed in post-classifiers (clean-up procedures), but on a pixel level rather than on an object level. Distance and direction are properties also arduous to utilize in classifiers, and rarely successfully used in remote sensing. An aberrant example is made by Gurney (1982), who identified clouds and cloud shadows in Landsat multispectral data.

Context

In the coming generation of image processors context will be much more involved than it is today. A knowledge-based system may include logics that are self-evident in manual interpretation. For example

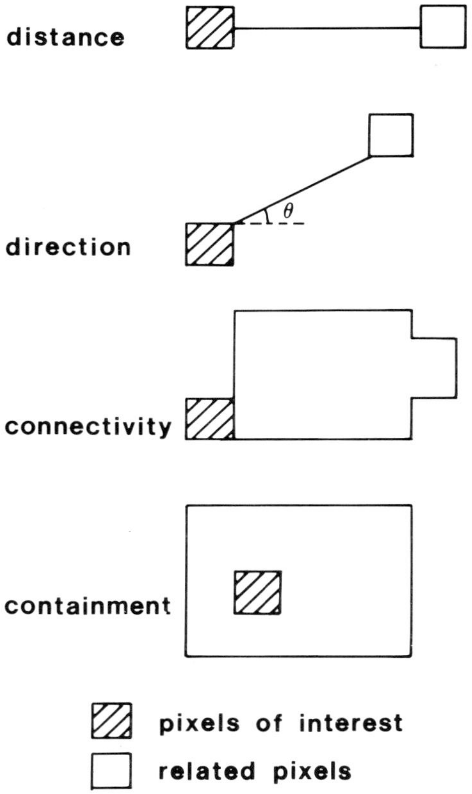

Figure 6.1 Spatial relationsships (after Gurney & Townshend 1983).

- no mature forest where a clear-cut was found last year;

- no clear-cuts in a city;

- no roads in a lake.

These classifiers also learn from the operator input or from results of earlier classifications.

When identifying villages in satellite data, it is difficult to use context information. Villages are spatially related to dirt roads, but these are seldom present in MSS data.

Possibly, the spatial relationship between the villages and their halos could be utilized. But, since no such intelligent image processor was available, context could not be utilized for automatic village discrimination, but certainly in the manual interpretation.

Texture

Together with object structure (as e.g. shape, which is discussed in Chapter 7), texture and context are characteristics always involved in manual image interpretation. To the human eye, texture is the apparent minute pattern of detail of a given area, and may be described in such terms as smooth, fine, rough and coarse (Hsu 1978).

The relationships between human texture perception and texture recognition by computers were ventilated and tested by Hsu & Burright (1980).

Texture is lively discussed and increasingly used in applied remote sensing. The program construction is more complex (for context and structure much more complex) than it is for per-point classifiers.

There are two main methods of implementing textural information into classifiers. One is to transform the image through textural filtering and simply use the output feature as additional input channel(s) into the classifier.

Another approach is a more complicated classifier, which automatically uses the spatial information in its algorithm. The latter is often referred to as 'contextual classifier', which is somewhat ambiguous, since it is dealing with pixels solely, and not objects. Saebø et al (1985) comprehensively described some of the existing methods, and also suggested a new set of contextual classifiers.

The texture analysis is exclusively two-dimensional (operating with only one input feature at a time), although it is possible to imagine an n-D analysis.

In most applications texture features are used to support the spectral classification, to achieve better classification results. Texture features are infrequently used to separate classes, which do not exist in the spectral domain, e.g. to separate cultivated areas (consisting of many, spectrally diverse fields) from areas with natural vegetation (Cf Olsson 1985).

6.2 Filters

The concept 'textural filtering' denotes here a micro-texture approach, where the centre pixel in a small window (e.g. 3 by 3 pixels) is exchanged by a value calculated from all pixels within the window. As the window moves over the image, a new feature is created.

Window size

The size of the window may be larger than 3 by 3 pixels, but the larger the window, the higher the risk of including pixels from another class. When classifying land cover types in Italy, Gurney & Townshend (1983) found that a 5 by 5 pixel window was superior to both 3 by 3 and 7 by 7 pixel windows. Based on his own experiences, Hsu (1978) suggested 3 by 3 pixel windows, since even 5 by 5 pixel windows have a substantial edge effect.

Edge-preservation

Edge effects can be avoided by edge-preserving filters. These commonly have a relative large window, but only use the group of the pixels that has the least standard deviation. There are two common means of discriminating the pixels to use:

1) Rotation, where the groups in each 'geographical' part of the window are tested for their standard deviation;

2) Sorting, where the pixels are sorted after DGL magnitude, and then divided into groups which are tested for their standard deviation.

Thus, only pixels affiliated with the same class are taken into consideration, and not pixels on the other side of the edge.

6.3 Image description

In analogy the with statistical description of spectral properties of an image or a class, one is interested in describing the textural properties with statistics, graphics and figures. Strahler & Woodcock (1984) suggested the use of variograms.

The 2-D variogram (Fig 6.2) is a histogram which has 'distance from window centre pixel' on the x-axis and 'local variance' on the y-axis. Theoretically, all pixels should be included when calculating the variogram, but a sufficient sample is enough.

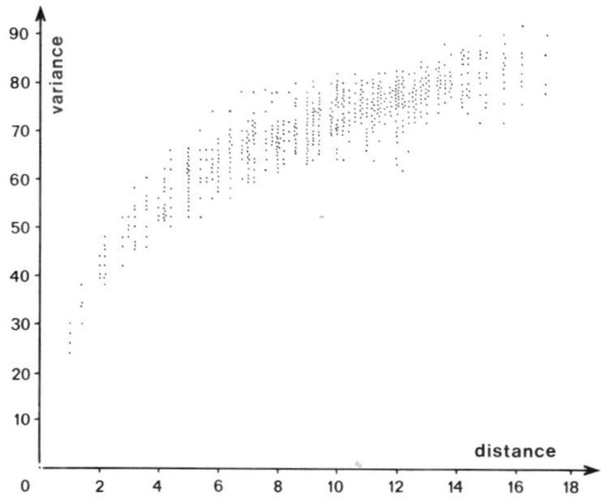

Figure 6.2 2-D "variogram" for the sub-area in Fig 6.4.

The pixels closest to the centre pixels are usually also closely correlated to the centre pixel, and the correlation decreases by distance. Normally, there is a break-point beyond which the variance curve flattens and does not increase. This is due to the structure of the objects in the image, or a training area, and their internal texture. Hence, the variogram may be used e.g. to determine a suitable window size.

The 3-D variogram also embodies the directional changes of variance. Figure 6.3 is an excellent example of how the variance changes in different manners, in different directions from the centre pixel. The two variograms describe the variance in the 512 by 512 pixel test sub-area of Fig 6.4.

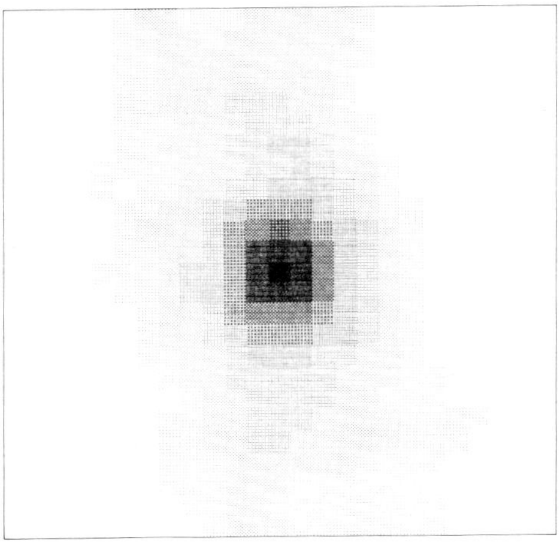

Figure 6.3 3-D variogram for the sub-area in Fig 6.4. Light grey tone corresponds to high average deviation from centre pixel.

The light elongated sand dunes and the dark interdunes in between can be clearly seen. Similarly, in the 3-D variogram the same pattern is visible. The correlation decreases faster perpendicularly to the dunes than along them, and it flattens at a distance of 5-6 pixels from the centre pixel.

The 2-D variogram concerns only the average variance versus distance. Regarding the appearance of the 3-D variogram it is easily understood that there is no obvious break-point where the variance curve flattens. Hence, there is no obvious window size that would 'fit' the objects of this image. It is more relevant to obtain variograms from training areas.

Tests revealed that variograms created from villages made no sense. The training areas were too small to discharge any fruitful information on internal village texture.

6.3 Review of some texture applications

Jaakkola et al (1984) found that the high resolution SPOT-data will benefit from classifiers which take spectral-spatial information into account.

Jensen & Toll (1982) experienced that texture + spectral data improved the classification of urban development, compared to spectral data alone.

Haralick & Shanmugam (1974) found that the overall classification accuracy increased from 74 to 83.5 % when including 8 textural features into the classification of 7 land cover types.

Frank (1984) tried to relate texture caused by high-frequency albedo variations to changes in erosion in semi-arid areas. However, the relationship was difficult to ascertain directly due to the low resolution of Landsat MSS data.

Shih & Schowengerdt (1983) added texture features to the spectral bands when classifying arid geomorphic surfaces with Landsat MSS data, and they improved the classification accuracy significantly.

Saxon (1984) found spatial associations between gradual gradients (=ecoclines) in multi-temporal transforms of Landsat imagery and

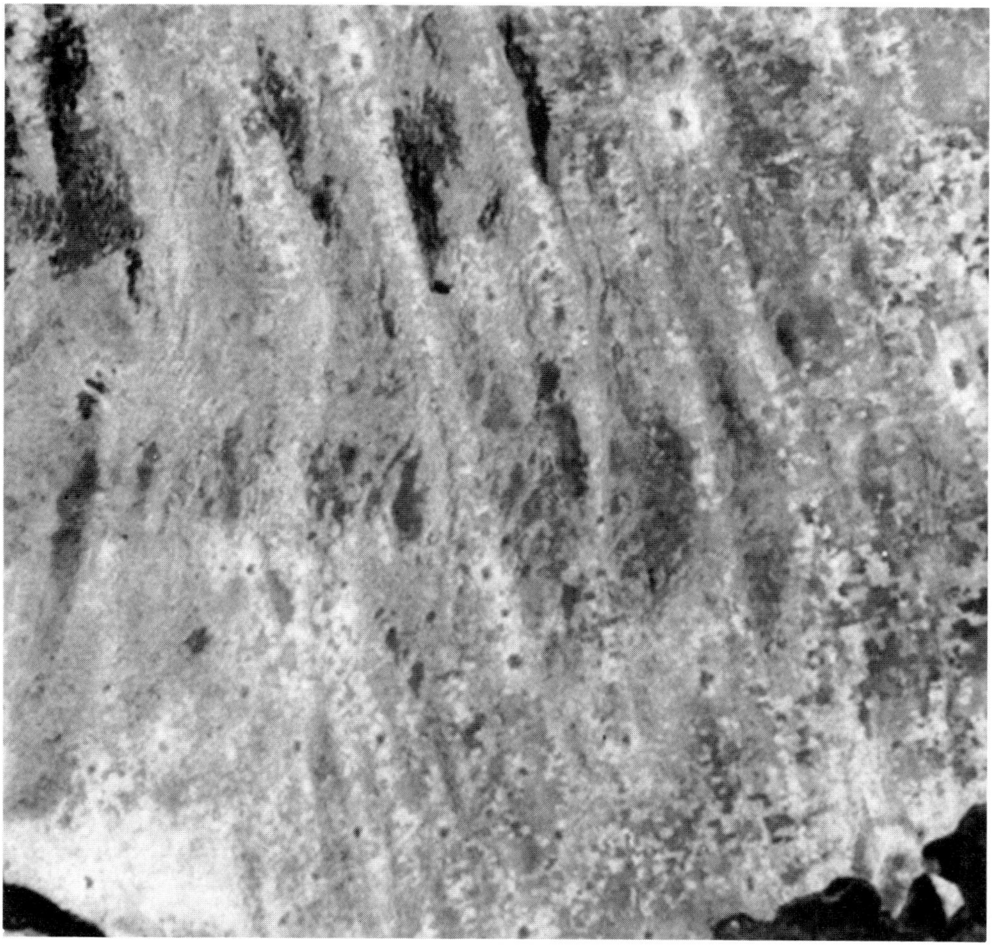

Figure 6.4 The sub-area used for all tests in Chapter 6. Compare with the variograms (Fig 6.2 and Fig 6.3) depicting the local variance in relation to centre pixel of a filter window. Effects similar to the ones in Fig 6.3 may be seen in Fig 6.5 a), where the Fourier power spectrum elucidates the dominating frequencies and their direction. This direction should theoretically be perpendicular to the one in Fig 6.3. Finally, Fig 6.4 might be compared to the textural transformations in Figs 6.7 a) through f).

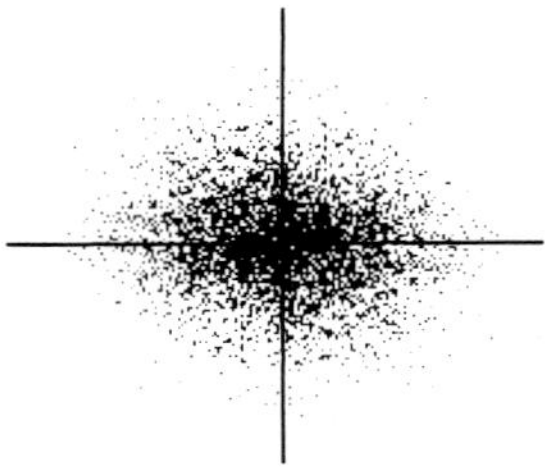

Figure 6.5 a) Fourier power spectrum of the test sub-area.

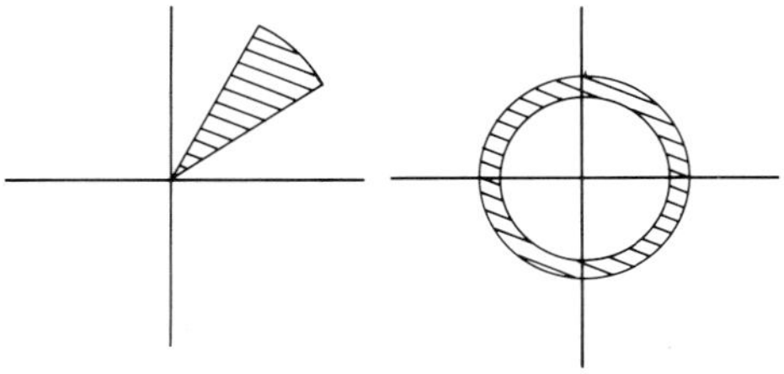

Figure 6.5 b) By integrating over the wedge in the power spectrum, information about directional biases can be derived. Similarly, the ring can give information on texture coarseness (after Goodenough & Goldberg 1979).

populations of vulnerable animals.

Alm (1985) added texture features to his classification of land-cover types using digitized aerial photographs, and noticed a considerable increase in classification accuarcy, for some of the classes, as compared with using spectral bands only.

Weszka et al (1976) compared a set of texture measures, among them the use of the Fourier power spectrum (Fig 6.5). Jensen (1979), inter alia, ventilated the spatial-dependency matrix. The latter, lively discussed at the end of the last decade, was abandoned in the early 1980s. The former is related to the methods adopted by the GOP system (see Knutsson & Granlund 1983 and Chapter 12), but is not further mentioned here.

6.4 Texture measures

Pattern recognition literature comprises innumerable texture transformations. Hsu (1978) and Haralick & Shanmugam (1974) list and depict a variety of them.

Besides the possibility to disjoin edge-preserving filters from other filters, the filtering-based texture measures may be divided into three comparison groups:

1) regarding central tendencies
   - e.g. standard deviation and skewness;

2) regarding centre pixel
   - relations between the centre pixel and other pixels within the window;

3) regarding inter-pixel variation
   - relations between neighbouring pixels within the window.

The two latter measures are often strongly correlated, while they theoretically differ from the former. Figure 6.6 illustrates the theoretical texture differences for cultivated and grazed areas in Kordofan. The measures using standard deviation (group 1) and inter-pixel variability (group 2) are compared.

Used measures

The present investigation involves 6 different texture measure, all applied to the test sub-area in Fig 6.4.

1) Standard deviation;

   A simple central tendency measure where the standard deviation was calculated in a 3 by 3 environment. In Fig 6.7a, the edge-enhancing effects of the standard deviation can be clearly seen, i.e. at the border between a vil-

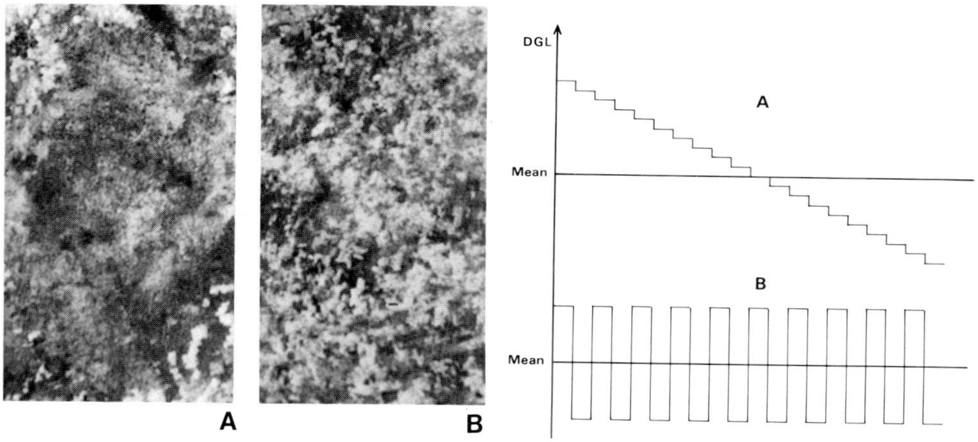

Figure 6.6 a) Texture differences between a grazed area (A) and a cultivated area (B). b) Schematic illustration of the one-dimensional DGL variations. Mean and standard deviation are equal for A and B, but the inter-pixel variability is considerably greater for B than for A (Stern 1984).

lage and its surroundings the deviation is high.

2) Frequency of deviation;

The number of pixels (within a 3 by 3 environment) deviating more than n DGLs from the centre pixel. In this case n was set to 3. This transformation (Fig 6.7b) is similar to the standard deviation filter, but has weaker extremes and comprises more noise.

3) Inter-pixel variability;

A measure, which is calculated with the formula

$$\frac{\sum |p_n - p_{ij}|}{\bar{p}} \qquad p_{ij} = \text{window pixels} \quad p_n = \text{neighbours}$$

and takes into account the differences between the closest neighbours in a 3 by 3 pixel window. The most striking appearance of this transformation (Fig 6.7c) is the extreme sensitivity to the banding in the image, else it is similar to the standard deviation filter. The banding (due to sensor differences) cannot be seen in the original image, although it is not destriped.

4) edge-preserving standard deviation;

Based on the sorting approach, this edge-preserving filter calculates the standard deviation of five sorted pixel values at a time. The lowest deviation is assigned to the centre pixel (Fig 6.7d).

5) homogeneity;

This filter (Jaakkola et al 1984) compares, within a 3 by 3 window, the pixel DGL with its eight surrounding pixel values (thus including pixels outside the window) by summing up all 72 values according to

$$\sum \frac{1}{1+(p_n - p_{ij})}$$

The filter transformation makes the output feature (Fig 6.7e) noisy, and the villages are difficult to identify.

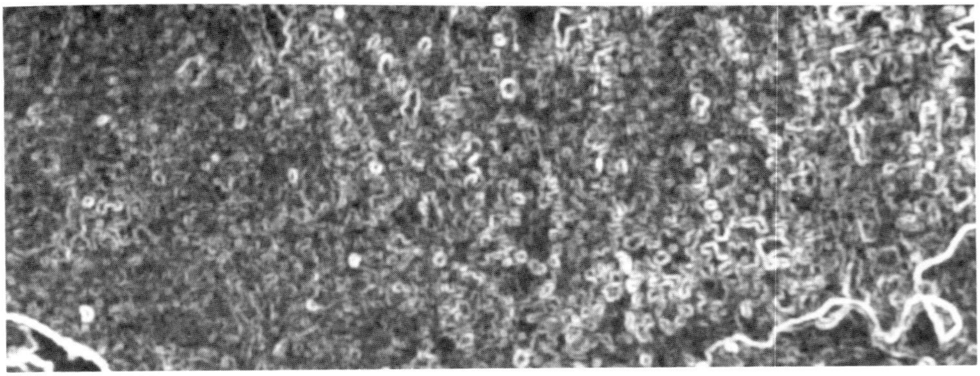

Figure 6.7 a) Effect of texture transform #1 on the sub-area (lower part)

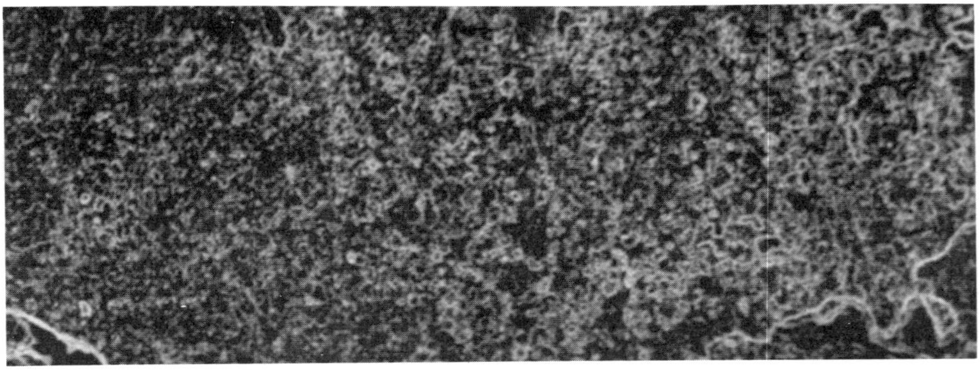

Figure 6.7 b) Effect of texture transform #2 on the sub-area (lower part)

Figure 6.7 c) Effect of texture transform #3 on the sub-area (lower part)

Figure 6.7 d) Effect of texture transform #4 on the sub-area (lower part)

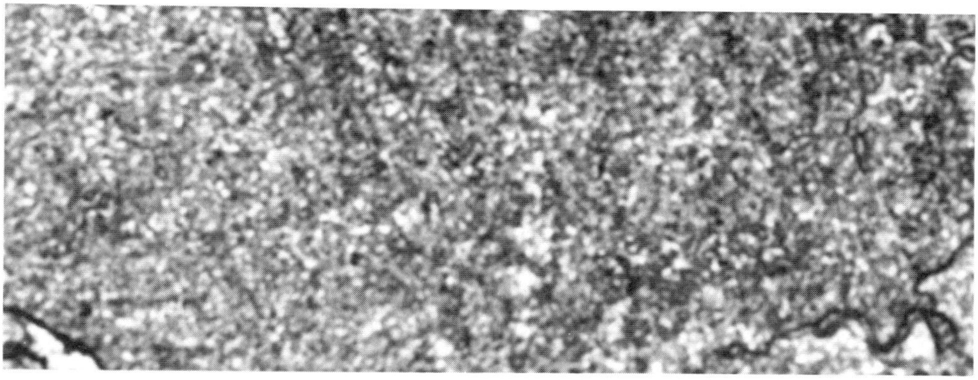

Figure 6.7 e) Effect of texture transform #5 on the sub-area (lower part)

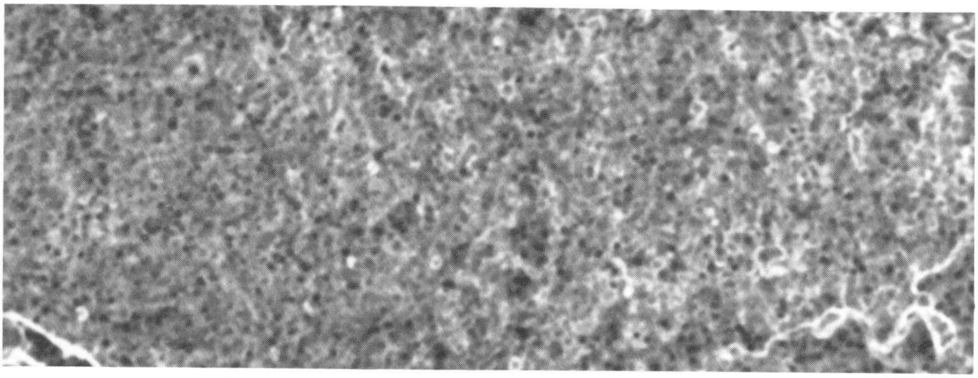

Figure 6.7 f) Effect of texture transform #6 on the sub-area (lower part)

6) 's-plane' texture;

Jaakkola et al (1984) proposed this filter, which in its first step is the same as # 4. In the second step, it assigns the mean of the five lowest values to the centre pixel. The output feature (Fig 6.7f) is very noisy and when negated, it is similar to Fig 6.7e.

6.5 Feature reduction

Jensen (1979) propounded that only a few texture features are necessary to characterize spatial relationships, e.g. a central tendency measure and a centre pixel variation measure.

Irons & Petersen (1981, p 365) presented a correlation matrix of 11 texture transforms. When examining the matrix carefully, it is obvious that there are only 2 useful transforms left when disregarding transforms with very high correlation. These were a variance measure and a contrast-to-centre pixel measure.

Input data

The first step in the feature reduction was to choose the spectral band to be used for the texture transformation. To test all spectral bands for all filters in a full classification test was unfeasible.

Instead, the approach of Chapter 5 was applied, i.e. test the separability for the most troublesome areas in the spectral classification. As an extra simplification, they were only applied to one of the texture transforms (Table VI.1).

The conclusion to draw from Table VI.1 is that the March image has larger differences in standard deviation for village training area and acacia training area than the September image has. This may suggest that the March image (dry season) also outrules the September image regarding manual interpretation.

Table VI.1 Separability of the 8 spectral bands when applied to a standard deviation filter. Only two classes, village and dense acacia senegal, were used, which means that the minimum separability is 50 %.

Band	Separability
Mar4	58.9 %
Mar5	57.7 %
Mar6	59.0 %
Mar7	59.0 %
Sep4	53.9 %
Sep5	53.3 %
Sep6	54.0 %
Sep7	55.1 %

Mar6 was selected as input spectral band, although the result of Table VI.1 does not prove the superiority of Mar6 for any of the texture transforms. But it suggests that the choise barely made any difference.

Transform selection

In exactly the same manner as when picking the spectral band, the transforms were chosen with a separability test (Table VI.2).

Table VI.2 Separability for textural transforms of Mar6, in accordance with Table VI.1

Feature	Separability
Orig Mar6	59.0 %
Texture #1	60.3 %
Texture #2	62.2 %
Texture #3	59.8 %
Texture #4	61.5 %
Texture #5	67.8 %
Texture #6	68.3 %

Table VI.3 Correlation between the texture features in the training areas.

	MSS6	tex1	tex2	tex3	tex4	tex5	tex6
MSS6	1.000						
Texture #1	0.054	1.000					
Texture #2	0.059	0.762	1.000				
Texture #3	0.186	0.734	0.603	1.000			
Texture #4	0.627	0.204	0.159	0.522	1.000		
Texture #5	0.004	-0.574	-0.593	-0.289	0.237	1.000	
Texture #6	0.060	0.736	0.665	0.629	0.203	-0.643	1.000

It seems that texture features are slightly more potent than spectral feature for discriminating villages from acacia senegal. However, the differences are small.

Feature reduction, based on feature correlation in the training areas, was carried out. Table VI.3 denotes the correlation coefficients for the 6 texture features and the original March MSS 6.

6.6 Conclusion

The separability differences between Mar6 and the texture features were not convincing, but a minor discriminating power cannot be ruled out.

Since the spectral features will be used in the combined automatic test, a feature strongly correlated to Mar6, is unlikely to add any information useful in a classification.

Similarly, the other features are strongly correlated to each other, which agrees with Jensen (1979), who stated that only a few texture transforms are needed to characterize a spatial relationship.

The test implies that one feature, Mar6 applied to the 's-plane' texture, was the most efficient, and was passed on to the combined automatic methods (Chapter 8).

7 morphometrical discrimination of villages

7.1 Background

As can be expected after reading Chapters 4 and 5, villages cannot be demarcated solely based on their spectral pattern. Village classifications result in a map with a number of village pixels, probably clustered into a number of objects, of which some are real villages and others are not.

The preceding chapters dealt with 'low-level' per-pixel classifiers and 'higher-level' classifiers regarding pixel neighbourhood. The present chapter will in some degree escalate the level of sophistication, and take a few steps into object classifiers.

Objects classification is not unknown in remote sensing applications, e.g. per-field classifiers are used for crop classification (Kauth et al 1977), but usually the objects (fields) are a priori determined from maps, and the classification variables are strictly spectral.

The present chapter discusses methods for automatic detection of objects (villages), automatic determination of object properties (size and shape) and automatic classification of objects based on object properties.

The measurement of shape, morphometrics, is a well-known tool in geography:

> "Shape is present in much implicit theory, running from the geomorphologist's concern for the lay of the land, to the urban geographer's classifications of the shape of cities. Since Christaller's time, work on the central place theory has been explicit enough to produce testable implications regarding the shapes of market areas" (Bunge 1962).

When Bunge wrote this, the gauge of geographical shapes had been restricted to maps and aerial photographs. Consequently, the methods were developed in accordance with the data, namely vector data, i.e. coordinate pairs describing the object. Raster data were indeed invented, but only in theory since few computers were powerful enough to handle large quantaties of raster data.

Still, most shape measures discussed for remote sensing are based on data in vector form, although remote sensing data are constituted of raster data. This implies a raster-vector conversion, which is not trivial to all image processing systems. Clarke (1984) stated that most of the existing vector shape measures may be converted into raster shape measures, which will speed up the computing.

Generally, metrics of properties involving or regarding the whole object, such as size, centroids and enclosing figures, are more easily calculated in raster data. Contrary, vector data are superior for metrics regarding the very outline of the objects, like perimeter and ruggedness. Joint vector/raster procedures may be the most sufficient means for remote sensing applications.

7.2 Related literature

Although morphometrics in general, as well as remote sensing, is a basic tool in geography, the combination of the two is cardinally discussed in theory, and seldom put into practice. Other branches of digital image analysis have overpassed geography long ago in this respect. In medical applications shape analysis is widely spread, and shape variables are essential in the realm of robot vision.

A few remote sensing applications may, however, be found in the literature. Size, which might be reckoned as a 'sub-unit' of shape, is the property most commonly used, e.g. by Davis and Peet (1977), who use minimum area of classified regions to clean up classification results.

Clarke (1984) used 14 known shape measurements to test their applicability to land-use classification of remotely sensed data. However, in Clarke's test the shapes were digitized from maps.

Lindell (1985), when studying water quality in lakes, used automatic measurements of lake area, shore line length and shore line development. The latter is related to the homogeneity measure described below.

7.3 Shape measures

Handbooks on general image processing and mathematical geography frequently report on different shape measurements. A few of them are depicted here. Basic definitions:

A = area of the object
P = perimeter
D = diameter of a circle with the same area (i.e. sqr(4A/π))
L = length of longest linear axis
C = a defined centroid
N = number of coordinate pairs
(x_i ,y_i) = i:th coordinate pair

The shape measures can be divided into different groups, depending on the property they exhibit, e.g. circularity or homogeneity. The described measures are all (if not otherwise stated) 'classical' metrics, and can be read about in more detail in, e.g. Austin (1984), Clarke (1984) or Castlemann (1979).

* Circularity, C_n

$$C_1 = \frac{P^2}{A}$$

where C_1 for a circle is 4π and other forms have higher values. C_1 is a scale dependent measure, why often a similar one is suggested;

$$C_2 = \frac{P}{P_C}$$

where P_C is the circumference of the circle with the same area. C_2 is 1 for circles and more for other objects.

* Elongation ratio, E_n

$$E_1 = \frac{D}{L}$$

which varies from 0 for a line to 1 for a circle.

* Rectangularity, R_n

$$R_1 = \frac{A}{A_r}$$

where A_r is the area of its minimum enclosing rectangle. R_1 is 1 for rectangular objects, $\pi/4$ for circles, and approaches zero for slender, curved objects.

$$R_2 = \frac{W_W}{W_L}$$

where W_W and W_L are the width and length of a minimum enclosing rectangle.

An extremely trivial rectangularity measurement, not found in the literature, is what may be called quadrangleness, which distinguishes between e.g. cultivated fields and complex non-anthropogenic objects. Quadrangleness is simply the number of vector pairs needed to describe the object. If this number is divided by four

$$R_3 = \frac{N}{4}$$

quadrangle-shaped fields will conclude in 1, triangles in 3/4, but more complex objects will have values higher than 1. Obviously, redundant coordinate pairs have to be sorted out by boundary smoothing, a classical procedure where every coordinate pair is compared to its neighbours.

For detection of anthropogenic element, it would also be possible to utilize measures that search for right angles (which is different from quadrangleness) or symmetry.

* Homogeneity, H_n

These are properties related to the local or total irregularity of the circumference, often the distance, d_i, from a centroid to the edge.

$$H_1 = \sum_{1}^{16} \frac{|d_i - \bar{d}|}{16}$$

where $\bar{d}$ is the average distance. d_i is measured in 16, or more, equally spaced radii from the centroid. H_1 records 0 for circles and higher values other shapes.

$$H_2 = \sum_{1}^{16} \frac{|d_i - d_{i-1}|}{16}$$

which is an inter-variability measure, which might distinguish between asteroidal objects with different number of vertices.

In raster data one may use the distance, d_j, from every pixel within the object to its nearest boundary pixel,

$$H_3 = \frac{N^3}{\sum_{1}^{n} d_j}$$

This measure does not require vector data, since it is equal to the IOD of the object transformed after the distance to boundary pixels (consult e.g. Castlemann 1979, p 326).

7.4 Other approaches to shape analysis

One approach is to inscribe an equilateral polygon and sum up the distances and squared distances between each coordinate pair and its neighbours, second- neighbours, third-neighbours, etc, until all the unique sums are found. For a polygon with N coordinate pairs N/2-1 distance sums and N/2-1 squared distance sums are needed,

$$S_N = \sum d_N \qquad d_N = \text{lag - n distance}$$

$$S_{N^2} = \sum d_N^2$$

According to Bunge (1962), each shape will enjoy a one-to-one correspondence with one of these sets of sums, which is an interesting observation, since it creates a unique possibility to classify shapes into well-known classes. His example is hexagonality, where a regular hexagon is

S_1 = 8.4 units (polygon sides),
S_2 = 13.2, S_{1^2} = 11.7, S_{2^2} = 31.7

and consequently hexagonality is the same figures within 'confidence intervals'

S_1 = 8.1-8.7, S_2 = 12.9-13.5,
S_{1^2} = 11.4-12.0 and S_{2^2} = 31.4-32.0.

Austin (1984) noted that this procedure involves easy computations. However, specifying the inscribed polygon is a rather cumbersome operation, and such a measure is merely a description of the inscribed polygon. Austin had only found one application of Bunge's method since 1962. The approach was not tested in the present investigation.

Another attractive means is explained by Hord (1982) among others. Hord was interested in shape only, and supressed size, orientation and translation with different kinds of mathematical image transforms. Mellin, Polar-Cartesian and Fourier transforms were performed successively, and a shape value was determined from the power spectrum of each object. The method, which is chiefly developed for shape comparison/matching, demands highly advanced image processing systems, and was not tested in the present investigation.

7.5 Defining villages in MSS data

To be able to expose the raster data villages to the shape measures, their boundaries were vectorized, and each boundary (set of coordinate pairs) had to be assigned to one and only one object.

The first problem stumbled over, is the definition of the object, and the second is to define its boundary.

Centroid

To be able to pick the object, one point close to the object centre is needed. This point, which may be selected manually (Chapter 9) or automatically from a classification result, is probably not a 'true' centroid in a mathematical sense, but ample to select an enclosing area around the object. For rural villages in the Kordofan, a window of 25 by 25 pixels is enough.

A classification result consisting of geographically separated segments might be automatically vectorized using e.g. soft-ware on the Ebba II system. Each segment is given coordinate pairs describing its boundary, and from these coordinates a 'centroid' is easily computed.

However, another approach was taken. The entire classified image was applied to a skeletonizing procedure (Rosenfeld & Kak 1976), where the boundary pixels are iteratively removed until only one pixel from each object remains (see Section 8.3). This pixel is then regarded as the 'centroid'.

Boundary

As seen in Fig 3.8, the outer limits of a village is not distinct on Landsat data, although they might be definite on aerial photographs. It is not possible to use classified regions as the true distribution of a village, since the level of the village pixels may vary from village to village. Hence, the classified region is often somewhat larger or smaller than the true distribution.

For villages, two main boundary selectors exist, both based on maximal gradient, but applied to different properties:

1) the gradient in the frequency histogram of the enclosing area of the object, portrayed in Fig 3.3;

2) the DGL gradient at the village fringe, between the village itself and its village halo.

The first approach is tasteful, since it needs much less computation than the second. However, when the method was tested, the failure percentage was much too high, although the method was excellent for 'standard' villages. The failures were due to other features than villages, within the enclosing area, disturbing the frequency histogram.

The second approach is awkward. The maximal spectral gradient is searched for in 16 directions from the picked 'centroid'. However, this approach needs a 'true centroid', which cannot be determined until the boundary we are looking for is known. An iterative process is therefore implied, which suggests a lot of computations, unless the picked 'centroid' is close to the 'true' one. Thus, a more plausible 'centroid' is defined, which might be called the spectral centre of gravity, SCG.

Spectral centre of gravity

Generally, the SCG should be a DGL-weighted geometrical centre of gravity including all pixels within the object, i.e.

$$X_{SCG} = \sum \frac{X_i \times W_{DGL}}{N}$$

$$Y_{SCG} = \sum \frac{Y_i \times W_{DGL}}{N}$$

where X_i = x-coordinate
Y_i = y-coordinate
i,j = pixel row, columns
W_{DGL} = weight for that specific digital grey level (must be different for dark and light objects)

However, in the present investigation, the object is not known, and pixels outside the objects will disturb if the whole enclosing area is used. Hence, only pixels within a circle with a diameter of 7 pixels around the picked 'centroid' are taken into account, and the SCG is calculated by using only pixel values less or equal to the minimum DGL+2

$$X = \frac{\sum_{1}^{3} \left[(4-m) \times \sum_{1}^{N_m} X_{DGL_{l+m}} \right]}{\sum_{1}^{3} (4-m) \times N_m}$$

$$Y = \frac{\sum_{1}^{3} \left[(4-m) \times \sum Y_{DGL_{l+m}} \right]}{\sum_{1}^{3} (4-m) \times N_m}$$

This concept, where only the darkest pixels in the village are regarded, is supported by the village structure, where the darkest pixels in the village compose the village centre. The SCG is a more 'true' centroid than the originally selected one.

An interesting footnote here is that Boyce and Clark (1964), from a methodological point of view, discussed the use of the functional centre, CBD, of a city (instead of a geometrical centroid) when applying radials to shape analysis of cities.

Gradient searching

Commencing from the SCG the maximum gradient is searched for, but only continuing until DGLs high enough are found, to be certain to be in the village halo. Otherwise, one might continue far away from the village where high gradients of foreign origin may appear.

The following gradient concept is used,

$$G_i = (DGL_{m_{i+1}} + DGL_{m_{i+2}}) - (DGL_{m_{i-1}} + DGL_{m_{i-2}})$$

where m_i is the ith pixel in that specific direction. Consequently the coordinate of the pixel with the maximal gradient, the boundary pixel, is

$$\left(X\max(G_i),\ Y\max(G_i)\right)$$

This operation is performed in 16 equally spaced directions (Fig 7.1). Column-wise, row-wise and diagonal directions are performed without problems, but the remaining eight directions are cumbersome, since they need approximations of the pixel to use (Fig 7.1). It can be done with some sophisticated interpolation method, but a simple nearest neighbour method seemed sufficient.

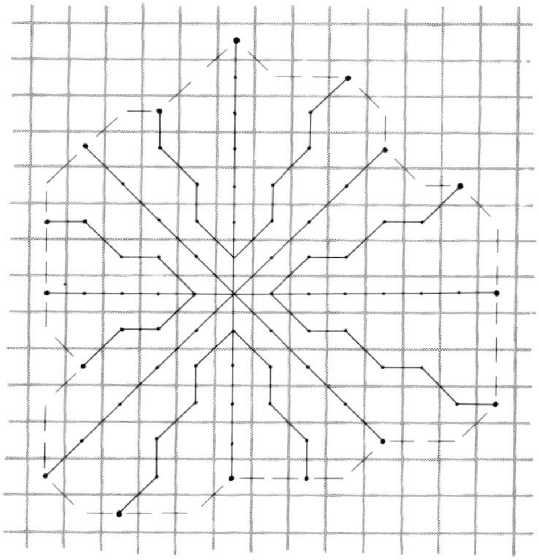

Figure 7.1 The search for maximal gradient is performed in 16 directions. The vectors connecting the 16 gradient pixels form the set of vectors for the shape analyses.

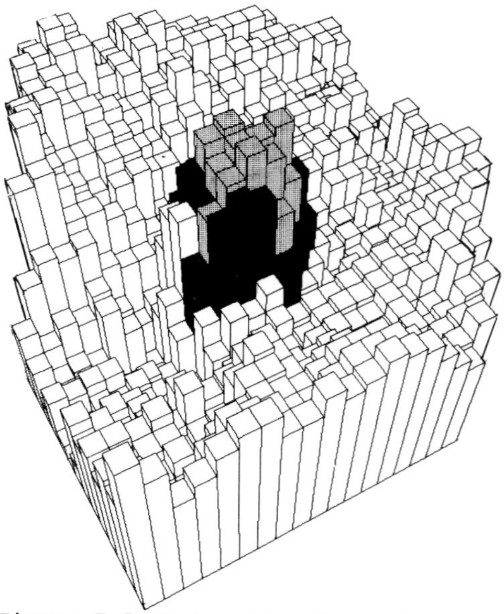

Figure 7.2 Automatic detection of village limits. Black bars represent the gradient pixels. The figure is reversed, high bars correspond to low DGLs, and the lowest bars are actually the light village halo.

The 16 pixels then form the set of coordinates necessary for the shape analysis. However, it is still questionable whether the 'true' centroid was used, i.e. that the 'correct' boundary was obtained. Thus, an iterative process starts, where a new centroid is calculated from the present boundary.

The new centroid entails a new boundary, a new centroid, etc, until the process does not change the centroid. However, the iteration process is aborted after 5 revolutions even if the centroid is exchanged, since the process might stall in jumps back and forth.

Problems of boundary demarcation

When an object needs more than 5 iterations of gradient search, it is indicated that the object structure is complicated, or that the gradient determination has failed, and the object is flagged for manual inspection.

It is difficult to analyse the accuracy of the method because of the psf and mixel problems, but a discussion is performed in Chapter 10, where the method is used for area determination. Nevertheless, it is obvious that 3 out of the 80 tested villages from the study area failed completely.

For these villages the maximum gradient in one of the 16 directions was placed far from the village fringe. Looking closely at the imagery it was obvious that the failed villages were located in an unrestful background, and thus, difficult to find in the March MSS 6 image.

These village gradients would not be found automatically with another algorithm, but it would be possible to flag for low values of max (G_i) and make a manual village delineation. The village would neither be found in any of the other MSS bands, but the problem would probably not arise in, inter alia, TM data.

7.6 Other methods

The boundary pixels might be collected from images transformed by a Sobel or a Roberts filter, which enhance gradients (see e.g. Schowengerdt 1983 for a theoretical discussion). The pixels with maximal DGL (= gradient) may then be regarded as boundary pixels, but these pixels values are calculated using gradients in several directions, while here only gradients radially from the 'centroid' is of interest. However, the difference might be small.

Few remote sensing applications found in the literature are similar to the present one. However, several medical applications are close. One interesting example is Bengtsson et al (1979), who segmented cells in images scanned in a microspectrophotometer. They compared cell distribution in images collected in different wave-lengths. The boundary of the cells were defined by looking for the maximum smoothness of DGL isolines around the center of gravity.

7.7 Tested shape measures and results

Four types of measures were tested in this investigation: C_1, E_1, H_1, H_2, and first applied to 25 geometrical figures and 79 villages digitized from aerial photographs (where the shapes were regarded as 'true').

Shapes in aerial photographs

The first problem to solve was which measure to utilize in a prospective shape classification. Table VII.1 describes the simple statistics for the villages and Table VII.2 depicts the correlation between the four employed measures.

There are high, positive or negative, correlations between all of the four tested shape measures. To

Table VII.1 Statistical description of shape measures applied to villages as measured from aerial photographs.

	Circularity	Elongation	Homogeneity 1	Homogeneity 2
Mean	0.63	0.24	4.28	4.25
Median	0.60	0.24	3.83	4.04
St deviation	0.13	0.03	1.34	1.75

Table VII.2 Correlation (r) matrix regarding the four used shape measures, using data from villages as measured from aerial photographs. Confidence intervals on the 95 % significance level are denoted within paranthesis.

	C_1	E_1	H_1
E_1	0.79 (0.68 - 0.86)		
H_1	-0.68 (-0.78 - -0.54)	-0.83 (-0.89 - -0.75)	
H_2	-0.73 (-0.82 - -0.61)	-0.83 (-0.89 - -0.75)	0.86 (0.79 - 0.91)

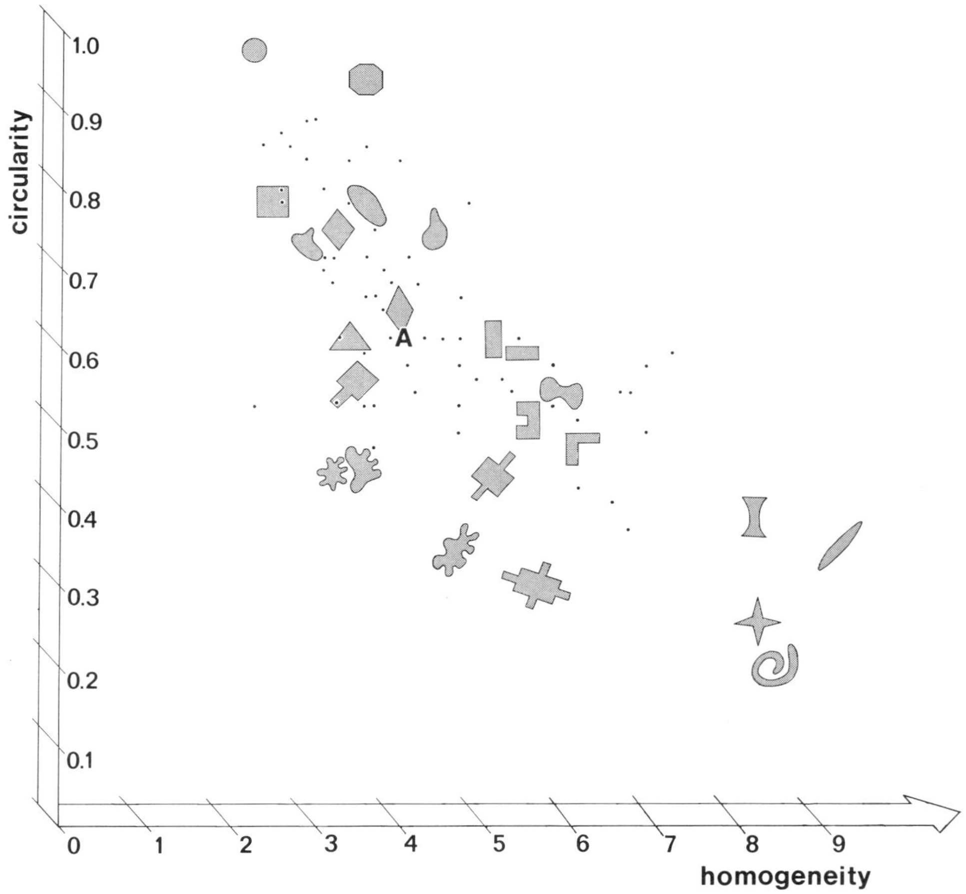

Figure 7.3 Bi-directional histogram representing the two shape measures Circularity and Homogeneity. Dots correspond to villages as measured from aerial photographs (A is the centre-of-gravity), and the test figures represent themselves.

extract all plausible information, the two first principal components may be calculated. However, for simplicity reasons, the two measures, C_1 and H_1, with the lowest mutual absolute correlation were used. To illustrate the effect of the two measures, they were plotted in a morphometric scattergram (Fig 7.3).

It is obvious from Fig 7.3 and Table VII.1 that village shape, as measured from aerial photographs, is a rather well-defined phenomenon. It varies in the 2-D morphometric domain within an area limited by a circle, a set-square and a 'baseball glove'. The centre-of-gravity for village shapes is close to the 'ace of diamonds'.

Shapes in MSS data

The fact that the 'true' shapes of the villages are characteric does not guarantee that the villages keep their shapes or shape measures when sensed by Landsat.

Keeping in mind the difference in appearance between objects non-zoomed and zoomed with cubic convolution (Chapter 3), both methods had to be tested. Table VII.3 depicts the correlation between shape metrics for villages as measured from aerial photographs versus convolved and non-convolved data, respectively.

Cubic convolved data were better correlated to aerial photograph data than were the non-convolved data, which implies the use of convolved data. However, the correlation coefficients are not convincing for the circularity and elongation measures.

The question remaining was the necessity of the Landsat shapes being well correlated to the 'true' shapes, or if they still were characteristic in relation to other objects discriminated in a spectral or textural classification. An answer is given in Chapter 8.

For the 'true' shapes C_1 and H_1 was the best two-measure combination.

Table VII.3 Correlation coefficients, r, regarding shape measures for villages in aerial photographs, raw Landsat MSS data and cubic convolved Landsat MSS data. Confidence intervals on the 95 % significance level are denoted within paranthesis.

Aerial photos	Non-convolved data	Convolved data
Circularity, C1	0.11 (-0.11 - 0.32)	0.40 (0.20 - 0.57)
Elongation, E1	0.04 (-0.18 - 0.26)	0.45 (0.26 - 0.61)
Homogeneity, H1	0.28 (0.06 - 0.47)	0.60 (0.44 - 0.72)
Homogeneity, H2	0.10 (-0.12 - 0.31)	0.67 (0.53 - 0.76)

Table VII.4 Correlation matrix regarding the four used shape measures, using data convolved MSS data. Confidence intervals on the 95 % significance level are denoted within paranthesis.

	Circularity	Elongation	Homogeneity 1
E1	0.76 (0.65 - 0.84)		
H1	-0.55 (-0.67 - -0.37)	-0.72 (-0.81 - -0.60)	
H2	-0.56 (-0.71 - -0.42)	-0.85 (-0.90 - -0.78)	0.85 (0.78 - 0.90)

Table VII.4 indicates the correlation coefficients for convolved MSS data.

Table VII.4, as well as Table VII.2, indicates that C_1, together with H_1 or H_2, was the best two-measure combination. Thus, it was convenient to put the MSS data values into the scattergram in Fig 7.3. In Fig 7.4 these values are shown in black, while the old Fig 7.3 is shown in grey.

From Fig 7.4 and the statistics of Table VII.5, it was noticeable that the cluster of villages has moved towards higher homogeneity values, while the circularity measure was stable.

Table VII.5 Statistical description of C_1 and H_1, when applied to villages as measured from Landsat MSS data (the figures within paranthesis are from Table VII.1 indicating data from aerial photographs).

	Circularity	Homogeneity
Mean	0.60 (0.63)	4.55 (4.28)
Median	0.62 (0.60)	4.45 (3.83)
St deviation	0.15 (0.13)	1.44 (1.34)

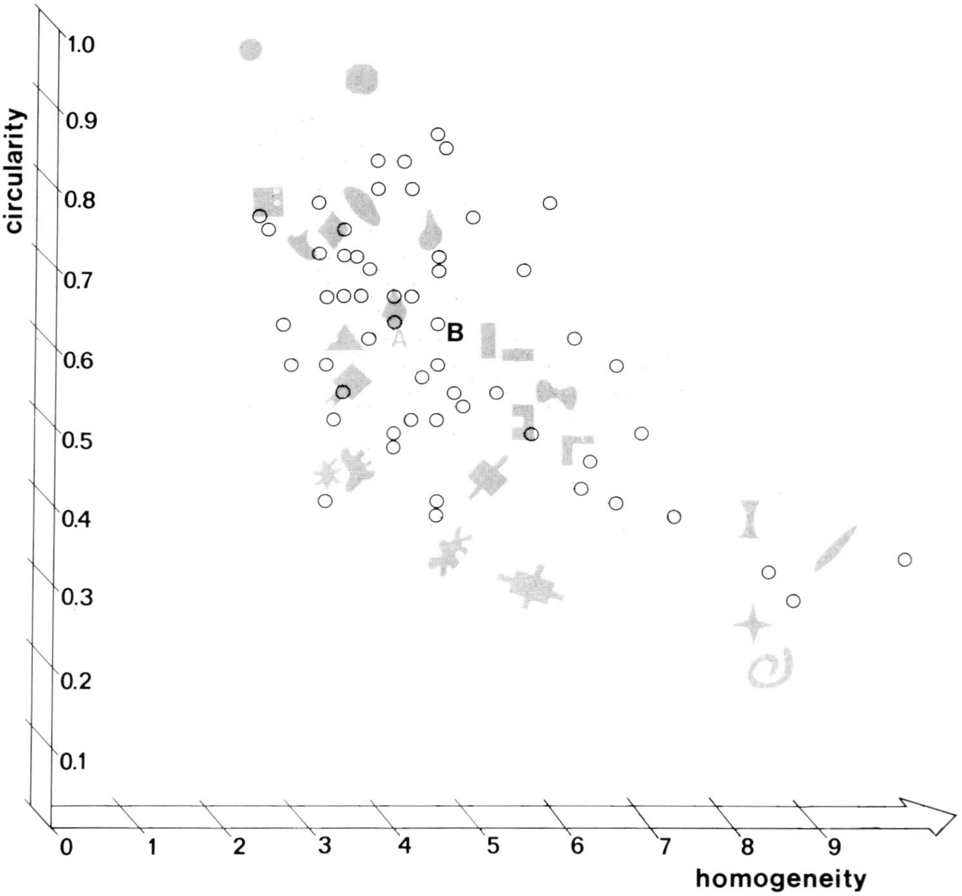

Figure 7.4 Bi-directional histogram representing the two shape measures Circularity and Homogeneity. Grey dots and test figures are the same as in Fig 7.3, and black circles are villages as measured in Landsat MSS data (B is the centre-of-gravity). The odd circle values in the lower right are also discussed in Fig 10.4.

7.8 Conclusions

Chapter 7 discussed shape measures and the behaviour of villages and test objects when applied to four theoretically chosen shape metrics. However, the villages were not tested against spectrally and texturally misclassified 'non-villages'. This test will be carried out in the next chapter, combined automatic methods.

The study indicated that the tested shape metrics were closely intercorrelated, i.e. they have similar sensitivity to different types of e.g. non-circularity. If morphometrics are applicable for village discrimination, probably two measures are enough. Circularity and Homogeneity #1 were passed on to the combined automatic methods.

Village shapes in Landsat data were much closer to village shapes in aerial photographs, if enlarged with cubic convolution. The correlation between MSS data and 'truth' was not convincing. But, looking at the scattergram (Fig 7.4), the villages moved within a rather limited area. Thus, despite the low correlation, a shape measure is possibly valuable as a village discriminator.

8 combined automatic methods

8.1 Background

In this chapter, the best methods suggested in Chapters 4 to 7 will be combined to the most efficient automatic procedure, which is compared to the most potent manual interpretation method in the next chapter.

As a recapitulation:

Chapter 4 suggested

1) to use a bi-temporal combination of four spectral bands, namely Sep7, Sep5, Mar7 and Mar4 for classification;

2) to test a classification using the four major principal components;

Chapter 5 suggested

1) the arithmetical combination Mar5/ (Mar4 + Mar5 - Mar6 + Mar7) as an additional band for classification;

Chapter 6 suggested

1) the textural feature 's-plane texture' as an additional feature for classification;

Chapter 7 suggested

1) a post-classification discrimination using the two shape measures 'circularity' and 'homogeneity' as discriminators.

The proposals in Chapters 4 to 6 regard features as input for classification, while Chapter 7 discuss post-classification procedures. Accordingly, the suggested feature combinations will be tested in a selected sub-area. The most efficient classification map will then, after a clean-up procedure, be applied to shape measurements.

8.2 Classifications with combined features

The proposed feature combinations

used in the classifications are

1. Sep7, Sep5, Mar7 and Mar4;
2. Sep7, Sep5, Mar7 and Mar4 + best texture feature;
3. Sep7, Sep5, Mar7 and Mar4 + best arithmetical feature;
4. Principal components 1, 2, 3 and 4.

Classification result

The three classifications were made with 3 standard deviations (SD), and all of the villages were identified in the chosen test sub-area. However, the maps inherited a substantial overclassification (the correct percentage village pixels is less than 1):

1. 15.65 %
2. 17.26 %
3. 13.80 %
4. 15.19 %

} pixels classified as villages in % of the total sub-area

Conclusion

The differences in over-classification were negligable, and since the village identification accuracy were equal for all classifications, it was possible to draw the conclusive inference that:

> Neither the arithmetical features, texture features nor the principal components did increase the classification accuracy. It is, however, important to note that this inference is not valid for discrimination of other kinds of objects or classes.

Further classification tests

The consequence of the drastic inference above was that only the spectral features were selected for further automatic processing. Based on the results of Chapter 4 also a classification with 1 SD was performed for the same sub-area.

The two classifications resulted in a village mapping accuracy of

- Mean + 1 SD 89.3 %
  (classified pixels 0.8 %)

- Mean + 3 SD 100.0 %
  (classified pixels 15.7 %)

An advantage of having two different classifications was that the benefits from both of them could be utilized, if combined with binary operators (see next section). These combinations were based on the experiences about the two classifications, namely

1) In the 3 SD map all villages were represented, some large villages had grown into other areas of classified pixels, while most small villages were still unconnected with other pixels;

2) In the 1 SD map all villages were unconnected, but some small villages were completely missing;

As mentioned in Chapter 4, when extracting features the mapping accuracy and the over-classification is important. But, when it comes to final village mapping, the classification maps must be turned into objects. And the mapping accuracy must be based on correctly identified villages, identified non-villages and omitted villages.

However, the actual map segments, or map objects, are of little interest, since only the coordinates of a centroid pixel are needed henceforth. The two maps (1 SD and 3 SD) were transformed into centroid coordinates.

The mapping accuracies (% identified villages) were still 89 % and 100 %, respectively, while the AND-OR values, due to substantial over-classification, were close to 0.1.

8.3 Binary operations with classification maps

A raster map comprising 'village

Figure 8.1a) 4- and 8-connectedness.

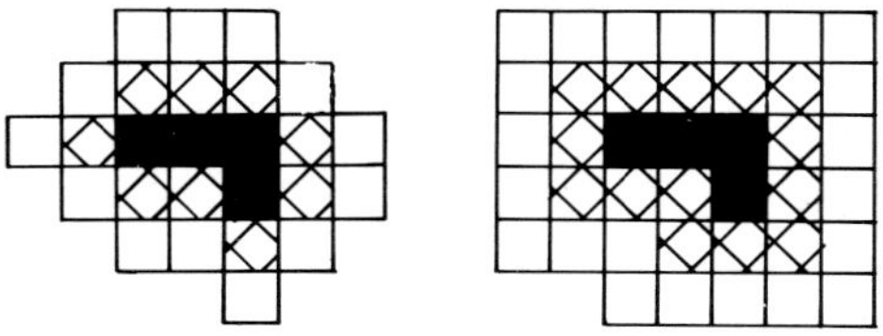

Figure 8.1b) Dilation employed twice with 4- and 8-connectedness, respectively.

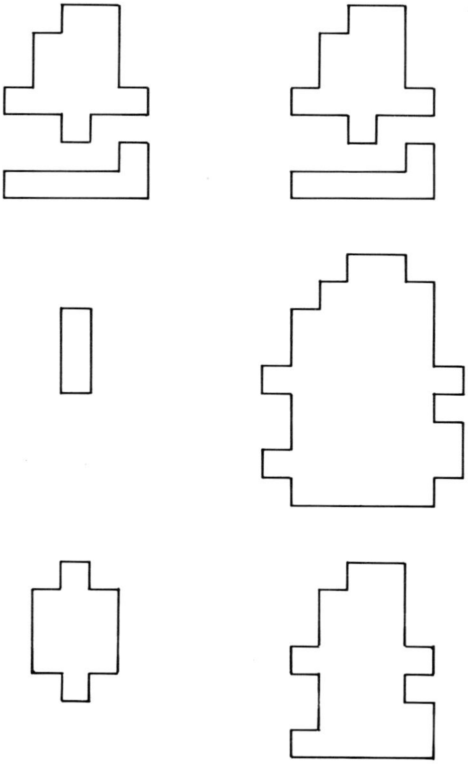

Figure 8.1c) 4-connectedness. To the left the test figure is affected by erosion followed by dilation, and to the right it is first dilated and then eroded (compiled from Context Vision 1985).

pixels' and 'non-village pixels' may be regarded and treated as a binary image, which has a row of properties allowing other operations than the grey-scale image does. One common pair of operators is erosion-dilation. For a more elaborated description, see Rosenfeld & Kak (1976), who, however, refer to these operators as propagation and shrinking.

Dilation is when a 1-pixel grows into its 0-pixel neighbours, and erosion is the opposite, when 0-pixels grow into their 1-pixel neighbours. When eroding and dilating are performed repeatedly, one may obtain results like filling holes in objects or separating connected objects (see Fig 8.1). Fig 8.1 also describes the difference between the use of 4- and 8-connectedness.

In the present study erosion-dilation operators were used to interconnect separated objects in the classification map. Erosion was also performed to find a centroid of each segment.

The Dipix system comprises binary-to-byte transformation where each output pixel value is the number of pixels in the segment to which the pixel is affiliated. There is a choice of 4- and 8-connectedness.

After dilating and eroding, the 3 SD map was applied to the binary-to-byte transformation. Then, it was density sliced at DGLs from 0 to 75, i.e. all large clusters disappeared. Simultaneously, the large villages were lost and just the small villages were still present together with small mis-classified clusters of pixels.

From the 1 SD map the large villages were picked together with some misclassified clusters. The two maps were logically OR-ed, which resulted in a map comprising less non-village clusters than before (Fig 8.4). All segments were then eroded to centroid pixels, used for further processing in the morphometric discriminator.

8.4 Morphometric discrimination

Complying the suggestions from Chapter 7, the centroid coordinates were used to initiate the gradient searching procedure, still in Mar6. The found boundary coordinates were then exposed to the Circularity and Homogeneity measures.

The Figures 8.2 and 8.3 indicate the windows which were selected around

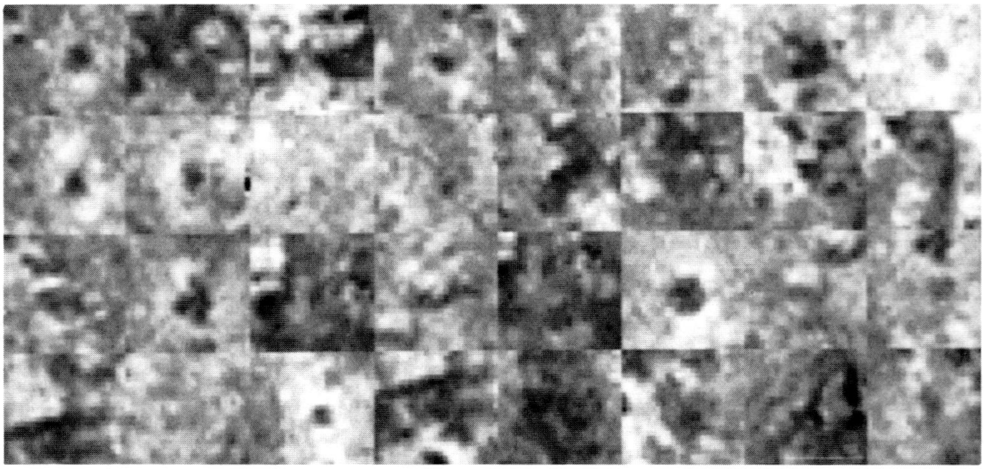

Figure 8.2 25 by 25 pixel windows around centroids selected with the best discrimination procedure. The correct villages are those similar to the ones in Fig 2.11.

Figure 8.3 Same windows with grey areas indicating the automatic gradient search. Note that also some of the correct villages have erroneous limits. Unfortunately, each window is turned 90 degrees by the computer program.

the centroids proposed by the best classification results. The human eye immediately spots some windows that definetely do not include villages, nevertheless, the were picked by the adopted automatic methods. In Fig 8.3 it may be noted that some gradient coordinates are found remote from an apparent circular object and, thus, overthrows the shape consistency.

8.5 Results and conclusions

Fig 8.4 denotes the AND-OR values for the original segmented map and the map when applied to shape analysis. Since a few villages were lost in the morphometric discriminator (probably due to gradient anomaly), Fig 8.4 also depicts the utmost possible AND-OR value, i.e. if no villages were lost. Hence, it indicates the minimum number of objects in the study sub-area that are spectrally, texturally and morphometrically confused with villages by the computer.

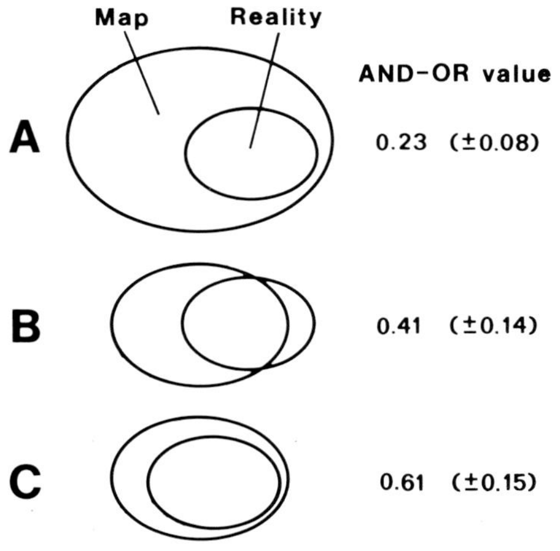

Figure 8.4 AND-OR values for combined automatic test. A) denotes the best classification, B) is also affected by morphometrical discrimination and C) is a hypothetical value assuming that no 'true' village was omitted by the shape measures.

The erosion-dilation process might be regarded as a scene-dependent ad-hoc procedure, difficult to transfer to classifications in other areas. It increased the AND-OR value from 0.1 to 0.23.

The morphometric procedure obviously increased the accuracy from 0.23 to 0.41 (a difference significant on the 95 % level), which indicates that shape measures were not useless for village discrimination, but neither were they potent enough per se to identify villages.

An AND-OR value of 0.41, which is low, was the ultimate accuracy for the chosen automatic approach, and a rather disappointing result after years of testing. The next chapter will unravel if the automatic procedure is competetive in comparison with manual interpretation.

9 manual identification of villages

Experiences accounted for in Chapters 1 to 3, were that the villages were rather easy to spot on aerial photographs and satellite images. But, as depicted in Chapters 4 through 7, it was difficult to automatically detect them. Hence, a manual discrimination of the settlements would possibly be preferable.

9.1 Basics of manual interpretation

Manual interpretation of Landsat MSS images is much like interpretation of aerial photographs, and the analogy increases with the resolution of satellite data. Thus, manual interpretation of TM and Spot data demand the techniques used for aerial photographs.

For a complete description of interpretation techniques, see Colwell et al (1983), and for a better understanding of the role of the human eye and the visual perception to manual image interpretation, consult e.g. Hubel & Wiesel (1979) and Estes et al (1983).

Typical characteristics used as clues for detection and identification of objects in a satellite image are colour, size, shape, texture, context and location (Fig 9.1).

For identification of Kordofan villages in MSS data these elements might be

- colour, size and shape of a settlement;
- existence, size and shape of a village halo;
- settlement location on sandy soils and at dune crests;
- similar distances between settlements;
- vegetation type or vegetation structure that implies deficiency of water and, hence, devoid of villages.

Man versus computer

The analysis procedures currently

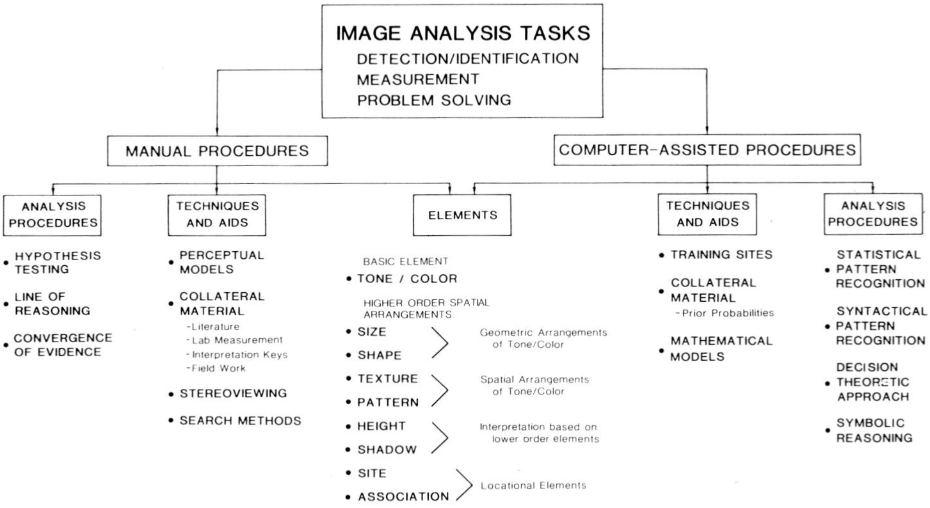

Figure 9.1 Image analysis tasks (from Estes et al 1983)

employed for automatic methods present a range directly analogous to human vision and visual perception, and their present state of development is extremely primitive when compared to human interpretation capabilities (Estes et al 1983).

Some of the disadvantages of today's image processing systems are clearly shown in the previous chapters. However, some computer properties are widely superior to that of the human brain:

- speed (sometimes);
- mathematical correctness when classifying objects, i.e. they can utilize smaller differences between classes;
- they can handle n-dimensional data (infinite number of bands,theoretically), while humans can only handle three bands displayed on the monitor or on a hard-copy image;
- they are not biased by intuition.

Comparisons

Although several authors have used both manual and computer-assisted methods in different studies of e.g. land cover types, there is an apparent lack of reported hard-core comparisons between the two approaches.

This might reflect a common opinion that the methods are not opposed to each other, but it is extremely difficult to find literature depicting a combination of the two methods. Either the analyst uses computers for a certain study or he uses equipment for manual interpretation.

Comparisons between manual and computer-assisted methods have, though, been performed by e.g. Helldén (1980). He concluded that approximately the same results can be achieved with either M-L classification or by manual interpretation of five land cover classes in Kenya.

The total mapping accuracy was slightly higher for the computer-assisted approach, but the difference was not significant at the 95 % level. Helldén also concluded that the manual approach was as fast, and somewhat less expensive than the computer approach was.

9.2 Review of village mapping literature

The investigation by Stern (1981) comprised a test comparing manual interpretation using standard colour diapositives from EROS Data Center and digital data displayed on a monitor.

Based on 20 interpretations (equalling 175 village samples), the results were unambiguous, the monitor approach was significantly better than the hard-copy approach. From the confusion matrices (unpublished) the AND-OR values may be calculated:

- Montior 0.54
- Hard-copy 0.39

These results were to be explained by higher geometric and radiometric resolution of digital data, and that the digital images could be specially enhanced for village mapping. No major differences were found between dry and wet season imagery, and multi-temporal imagery was not tested.

Reining (1980), when mapping villages from Landsat hard-copy imagery in Upper Volta and Niger (which have the same type of ecosystem and similar settlement patterns as Kordofan), found that:

- settlements larger than 200 m in diameter can be seen in the imagery, and that the size is measureable at the scale of 1:1,000,000;
- the likelihood is extremely good that the spatial extent and reflectance characteristics can be related to population densities;
- the dry-season imagery is much better than rainy-season imagery for determining the location of villages;
- repetetive imagery is necessary for adequate interpretation, because of the changes in appearance during the dry and the wet season;
- shape and location are relevant in making judgement on villages.

9.3 Identification procedures

Methodology

Considering the results found by Stern (1981), the manual identification of settlements was entirely conducted with digital data displayed on colour monitors. The interpretors were selected among the staff at the Remote Sensing Laboratory, University of Lund and at the Satellite Image Corporation, Kiruna.

The interpretors had different levels of experience in manual image interpretation, but they all got a short instruction and training (15 minutes) in village mapping ahead of the tests. Each of them interpreted two different band combinations and 2-3 sub-areas from each combination, but in different orders to avoid biases from increased knowledge about the area.

To utilize the advantages of digital image processing all the village identification tests were made in contrast stretch mode, i.e. the mapper had the opportunity to modify contrast and intensity of the image, and also to toggle between colour and B/W mode.

Data

The traditional manner of selecting Landsat MSS bands was used for the ordinary dry and wet season imagery (Table IX.1).

Table IX.1 The data sets (and their respective colour layers) for the tests of manual identification of villages.

Features	Red	Green	Blue
Wet season	Sep7	Sep5	Sep4
Dry season	Mar7	Mar5	Mar4
Bi-temporal	Sep7	Mar7	Mar4
Principal comp	PC 2	PC 1	PC 4

From the six most powerful bi-temporal 3-band combinations found in Chapter 4, one combination was subjectively chosen for the manual identification test. To the human eye, there seemed to be two groups, perceptionally close, among these combinations.

Both groups always comprised Mar4, the former compounded it with combinations of bands 6 and 7, while the latter preferred Sep5. After thorough manual inspection, a delegate from the first group was picked out (Table IX.1).

The fourth data set was chosen directly from the first four principal components of the K-L transformation. Components one and two comprised most of the information, whereas the other components inherited much of the noise. In addition to components one and two, also number four was selected (Table IX.1), since it seemed to capture the village halos.

The four images used for manual interpretation are displayed in Fig 9.2, although the reproduction of them here is of much lower quality than the monitor allowed.

Figure 9.2 The four images described in Table IX.1. a) wet season, b) dry season, c) bi-temporal and d) principal components. Compare with Fig 2.1.

9.4 Identification results

The results of the four identification tests are presented in Fig 9.3, where the AND-OR values are elucidated with Venn diagrams containing subset ellipses relative to the sizes of MAP and REALITY. Hence, the over- and under-estimations of villages may also be regarded.

The AND-OR values are denoted with their confidence intervals at the 95% level, and Table IX.2 denotes the significance of mapping accuracy differences.

Table IX.2 Significance matrix for AND-OR value differences of the four identification tests, based on T-test (95 % confidence level).

	Wet	Dry	Bi
Dry season	yes		
Bi-temporal	yes	no	
Principal comp	yes	yes	no

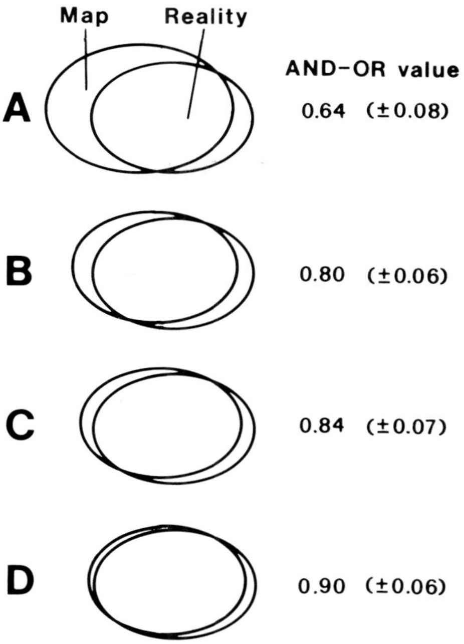

Figure 9.3 AND-OR values for of a) wet season, b) dry season, c) bi-temporal and d) princ. components.

9.5 Conclusions

Agreeing with Chapter 4, it was significantly shown that the use of multi-temporal data were superior to single-season data. However, it was more uncertain whether to use the principal component features or not, despite the clear appearance of villages in Fig 9.2.

It must also be remembered that the interpretation of principal components is strictly scene-dependent, i.e. the experiences of object colours etc, cannot be transferred from one image to another. This problem might to some degree be reduced by computing the principal axes from selected areas, or the use of canonical transformation instead of K-L transformation (see Curran 1985).

In accordance with Reining (1980), it was also observed that dry season imagery is superior to wet season imagery.

No obvoius trend of over/under-estimation was found. However, there seemed to be a slight predominance of over-estimation for the single-season images and a slight under-estimation for the multi-season imagery.

When it comes to population estimates, also the sizes of over/under-classified objects are important. This matter is further elaborated in the following chapter.

The most important conclusion of the study is, however, the indisputable superiority of the manual methods compared with the automatic ones. Consequently, the manual methods chosen will be the ones tested for the population mapping of a census district outside the study area (Chapter 11). But, prior to that test, the relationships between settlement size and population size have to be investigated and determined.

10 size/population relationships

10.1 Data needs

So far, all discussions have considered different ways to identify a village, i.e. midway towards the objective to determine the number of inhabitants in an identified village.

The ultimate goal requires data to correlate village sizes with population data and to formulate regression equations. The 'ground truth' data for these analyses originate from a couple of sources,

- field work made by our research group;
- field work made by other researchers;
- auxiliary population data.

The variety of data sources is caused by the problem of acquiring this kind of data at all. Data from different sources also considerably increase the problem of determining the reliability of the data, and thus the reliability of the overall analyses.

The data needed should be obtained as close as possible to the date of the Landsat data acquisition, i.e. March and September 1979. This was the latest acquired Landsat data until February 1985 over this particular area! However, 1979 was the year before this investigation first germinated, which hazards the fidelity of the statistical relations.

The most relevant parameter to this study is the 'de jure' data on

- 'total number of inhabitants in the village'

but ancillary parameters might be of interest, e.g.

- 'tribe' or 'dominating livelihood system' for stratification reasons;

- 'number of households' for crowding index;

- 'number of huts and houses' for spectral studies;

- 'type of market' for discussions on central places;

- 'type or size of water point' for the same reason and for the relation to size of village halo.

10.2 Field surveys

Many of the problems experienced and depicted below are well-known problems commonly discovered in total censuses or sample surveys. Some of the problems were also discussed in the introduction.

Population data were mainly collected in the study area between Er Rahad and Umm Ruwaba, because of the possibilities of studying thoroughly the visited villages on aerial photographs.

Members of the research team from the Department of Physical Geography, University of Lund, did field work in Kordofan in February-March 1980, February-March 1982, February-March 1983 and February-March 1984.

During most of the field work, where the main scope was other than population studies, only the 'total number' variable was collected, and only in a limited number of villages.

Data collection technique

In these cases, the field sampling vehicle stopped for a few minutes in the village or close to the village. Any person present was asked about the number of inhabitants, and sometimes also about other variables. For more detailed information on the field sampling technique used, consult Olsson & Stern (1981).

For several reasons, this is a very dubious means of data collection. The reliability of the answers obtained is proportional to the confidence in the questioner, which cannot be solidly established after a couple of minutes.

Watson & Tippett (1975) used veterinarians for a ground survey of household densities in the Sudan. The advantages with veterinarians handling the questioning was that most poeple were inviting and relatively frank to veterinary personnel.

During a field work in the town of Bara in Kordofan including interviews of 10% of the households, Hammer-Digernes (1977) experienced that some poeple believed that the Evil Eye might be thrown upon the family if the number of children is made known to a stranger.

In many villages the inhabitants refused to answer questions and directed to the chief (sheikh). In cases where answers were given, many ambiguous figures were obtained, and when asking several persons, the answers differed substantially. This is not due to intentional lies, but rather to recall lapse or lack of knowledge.

Hence, it is much more reliable to spend a longer period of time in each village, and perferably together with the sheikh. However, it also implies much more time required for the field work for the same amount of data. A well-known trade-off between quality and quantity.

In a special field survey for collection of population data, more time was spent in each village. In most of the instances, the data were collected during a tea session with the sheikh (Fig 10.1). The following variables were collected:

Questions

- name of village
- tribe
- number of households
- number of huts
- number of houses

Figure 10.1 Field interview and tea session with sheikh.

- present number of inhabitants
- number of inhabitants 1977
- number of market days/week
- distance to water point

Observations

- size of village halo
- soil type
- dominating land use
- settlement density

A similar field survey was carried out by personnel from the Department of Geography, University of Khartoum, under the direction of Professor Mustafa Khogali. Together with his students, he collected data on household etc, in some villages in the study area. The same variables as above were collected.

It was extremely difficult, without any auxiliary comparison data, to determine the reliability of these data. Especially variables like 'number of inhabitants 1977' are probably very unreliable.

Additional problems

Not only are those population figures obtained in the field somewhat unreliable, but also collected several years too late, for the comparison with the aerial photographs and the Landsat data.

Another problem could have been people living in the area but outside the villages. However, in the study area they are negligable. Most nomads and semi-nomads have a hut built in some village, and they were thus 'counted'. The same goes for seasonal workers that live in the Gezira area south of Khartoum a couple of months per year.

It is impossible to determine the influence of temporarily built huts, which might be subsistent during the dry season. When studying changes in population sizes, it is recommended to use images from the same season, to decrease the differences in seasonal dwellings.

Actually, the latter two problems regard more how migrators are counted in stated population figures than how they are 'counted' in the size measures.

10.3 Auxiliary population data

One of the non-governmental population data sources available was the Sudan National Livestock Census (Watson et al 1976). After light aircraft sample surveys and field surveys, they constructed, inter alia, maps on population densities for different strata all over the Sudan.

A smaller source was the 'Land and Water Use Survey in Kordofan' made by the consultant company Doxiadis Associates (Doxiadis 1966). They mapped inhabited land from aerial photographs and added population figures from the 1955/56 National Census to make population density maps (Cf Chapter 11).

Sudan Population Censuses

Problems arose when using data from the censuses. Data from the 1973 census was only published for large aggregates. From the 1983 census, which has yet only resulted in very summarized reports, it was possible to obtain preliminary data together with the 1:100,000 scale census maps.

These maps, as mentioned in Chapter 1, were enlarged from the topographical maps and included manually corrections of villages. The preliminary data comprised manually copied lists (translated into English) on population data regarding Kordofan.

Census data in the Sudan exist on different levels:

- nation
- province
- rural or town council
- village council
- village

Unfortunately, only data aggregated to village council level (2-3 villages) could be obtained. Hence, the data were worthless for detailed comparative studies of population and village size, but nevertheless, used for general comparisons in Chapter 11.

10.4 Areal measurements of villages

Size measurements could be carried out in raster data, simply by counting the number of pixels within the determined village boundaries. But, since the village boundary determination is made in vector data (Chapter 7), the used formula is

$$\text{AREA} = \frac{\left|\sum(x_i \times y_{i+1}) - \sum(x_{i+1} \times y_i)\right|}{2}$$

where X_i, Y_i are coordinate pairs in a known unit.

Since the spectral properties of the villages and its surroundings are rather stable and well-known, it is possible to use fractions of the edge mixels for a more accurate size measurement. The percentage 'village' in an edge mixel can be calculated through

$$\frac{(A - e)}{(A - B)} \times 100$$

where e = edge pixel DGL
A = village pixel DGL
B = halo pixel DGL.

A and B have either to be defined by an average value, calculated in an environment of 'true' village and halo pixels, or calculated perpendicular to the edge direction through:

max (Gx, Gy, Gu, Gv)

where Gx through Gv are the grey level gradients in the two vertical and the two diagonal directions. However, considering the overall accuracy of the approach and that the edge mixels probably affect the measurements both positively and negatively, the fraction pixel calculations were neglected.

10.5 Theoretical relations

Logarithmic relations

The relation between the size of a settlement and its population has been discussed by several authors, but a more mathematically elaborated approach is generally credited Nordbeck (1971). The law of urban allometric growth was formulated as

$$y = a \times x^b$$

Intelligently, Nordbeck discussed the dimensionality in the formula.

Population is intuitively considered a volume of the dimension 3, and the area of the settlement is evidently 2-dimensional. Hence, the constant b ought to be 2/3.

In his Swedish test areas, b proved to be 0.664 and 0.650, respectively, which is very close to 2/3 (Nordbeck 1971).

Welch (1980) swapped a and b in Nordbeck's formula and used

$$\log \text{POPULATION} = \log a + b \times \log \text{AREA}$$

which is the same formula, except for the values of a and b.

Inspired by Nordbeck, Wellar (1969) attempted to predict urban populations of American cities. His conclusion was that built-up areal extent and population were not necessarily closely interrelated.

Anderson & Anderson (1973) had problems when delimiting some Kansas communities with semi-automatic and automatic methods. They also obtained a combined error rate of only 6.9 %, while some individual communities erred +- 40 %.

Ogrosky (1975) found a high correlation (r = 0.982) between population and city area (both variables logged), when studying 18 American cities.

Linear relations

An often used relation between size and population is the simple linear regression:

$$\text{POPULATION} = a \times \text{AREA}$$

Foresti & de Mondonca (1974) found a good linear association between settlement sizes, as measured from MSS data and population, except for the largest cities.

Anderson & Anderson (1973) experienced very small differences between the use of Nordbeck's formula and the linear approach.

Stratification

Regarding the internal village structure depicted in Chapter 2, it is important to know the deviation of population densities among the villages. It is easy to guess that different tribes construct their villages in different manners, and hence, have different population densities.

Generally speaking, this is probably true. However, in the parts of Kordofan visited during the field work and studied from aerial photographs, there was no obvious deviation from the mean density. No relation, neither between structure and settlement size, tribes nor livelihood system, was found. But, when using the method in future population counts over larger areas, it is extremely important to make a comprehensive stratification.

Reining (1980) found, when mapping villages in West Africa, that the population density in villages affiliated to different tribes varied substantially. The Hausa tribe comprised 140 persons/ha, Songhai 174 persons/ha and Mossi 302 persons/ha. Cf the figures in Chapter 11.

10.6 Statistical relations

The uncertainty of the data allows only the conclusions to be qualified guesses, and the relationships described are merely illustrations of what could have been done.

Data

When some villages had been sorted out, due to location or identification problems, only 50 villages remained with data concerning size and number of inhabitants. These data were utilized to establish the size/inhabitant relationship, which were also used for the population mapping in Chapter 11.

The uncertainty mentioned concerns

- recall lapse regarding number of inhabitants;
- unsuccessful village area measurements.

Therefore, it is wise to introduce some logical reasoning and statistical analyses including data with higher reliability. Such data might be the size of the villages, as measured from aerial photographs.

From these photos also the number of households was counted, with a reasonable level of accuracy (unknown, but probable). Hence, a crowding index comparison was possible.

Another parameter comparable to 'number of inhabitants' is 'number of inhabitants 1977', since it is an answer given by the same person.

Reliability of inhabitant number

For 22 of the 50 village data was collected on

- number of inhabitants 1982 as stated by respondent;
- number of households 1982 as stated by respondent;
- number of households 1977 as measured from aerial photographs;
- size of settlement 1977 as measured from aerial photographs;

and 45 of the 50 villages comprised data on

- number of inhabitants 1977 as stated by respondent.

The first four variables were collected during the two special made field surveys, where comparatively more time is spent in each village. The statistical analysis regarding correlation between the variables resulted in Table X.1.

Presupposing that the size measures and the households counts made from the aerial photos are 'ground truth', the implication (but not statistically proven) is

> The rather high correlation between size and households as measured from aerial photographs indicates a homogeneous internal village structure, and suggests that a stratification is not necessary for the present test area.

This specific correlation was also determined for a larger sample (85 villages) of the study area, and with a coefficient of 0.942. Hence, the correlation between number of households and size is solid.

Crowding index

The crowding index, or number of persons per household, varies hea-

Table X.1 Correlation coefficients (N.B. calculated on an insufficient number of cases) regarding village sizes and number of households and inhabitants, based on 22 villages in the test area.

	Inhab 1982	Househ 1982	Size 1977
Households 1982	0.95 (0.87-0.98)		
Size 1977	0.89 (0.78-0.96)	0.88 (0.72-0.95)	
Households 1977	0.96 (0.89-0.98)	0.94 (0.87-0.98)	0.90 (0.77-0.96)

vily from household to household (Fig 2.8), but the averages for the villages in the same stratum, are likely to be stable. In accordance with Section 2.6, figures in the range 5 to 8 are anticipated.

The histogram in Fig 10.2 denotes the deviation of answers for the 22 villages. Some of the stated population numbers, or households numbers, are most probably wrong.

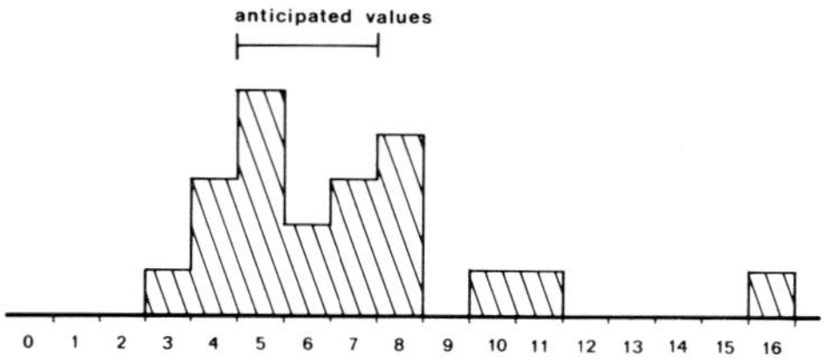

Figure 10.2 Crowding index values from 22 villages.

Growth rate

The rather strong correlations between the different figures from 1977 and the different figures from 1982 indicates a stable growth rate in the villages. But correlation coefficients do not tell the whole truth. Also growth rates for all villages with 1977 and 1982 population data were calculated. Fig 10.3 illustrates the distribution of these growth rates in a histogram.

The average growth rate for the villages, based on reported number of inhabitants, were 5.31 % (standard deviation 1.93). This number is considerably higher than the rates given by the censuses 1955/56, 1973 and 1983 (see Table I.2 and I.3). For the entire Kordofan these state 1.8 % and 2.5 %, respectively.

The discrepancy is instinctively credited recall lapse regarding the

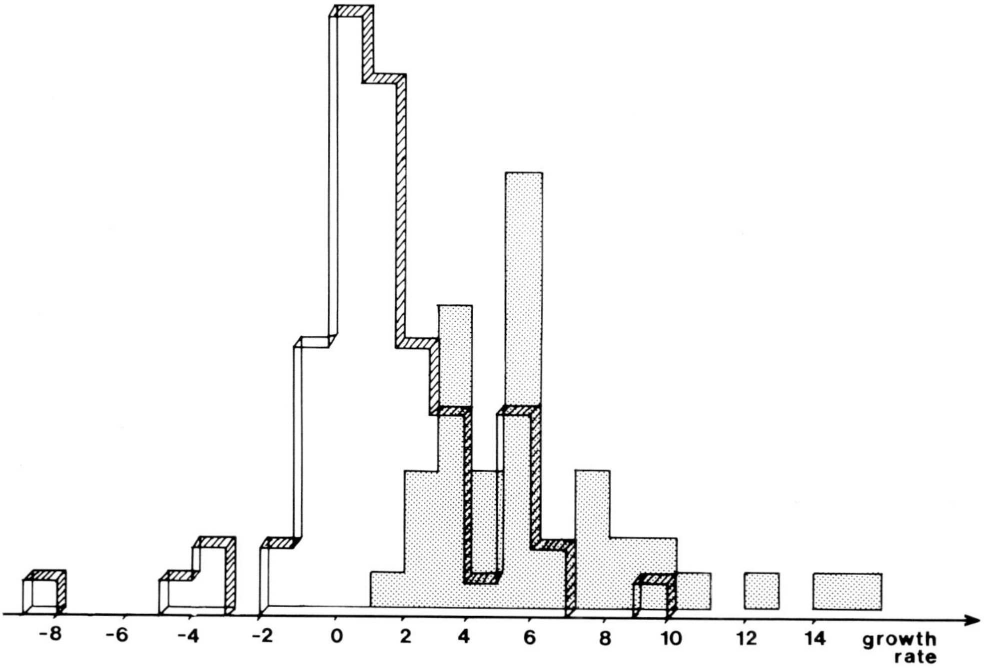

Figure 10.3 Growth rate values. Transparent areas are calculated from measures in aerial photographs from 1962 and satellite imagery from 1979. Hatched areas are calculated from stated inhabitant numbers.

number of population 1977. As a check, a test was performed using data from aerial photos from 1962 and satellite data from 1973. The result is seen in Fig 10.3, which indicates an apparent deviation from the average, but a magnitude much closer to the official census figures.

A common preassumption is that the growth rate increases with settlement size, i.e. that bigger villages grow faster than small ones. A quick test, which achieved a correlation coefficient of -0.161, indicated no such association.

Reliability of size calculation

Similar to the fidelity test of inhabitant data, the satellite size data may be tested against data emerging from the aerial photographs. 66 villages in the study area were included. Those villages consisting of three pixels or less were left out, since they could not be unequivocally determined in the Landsat data.

The relationship between size as measured from Landsat data and aerial photos, respectively is shown in Fig 10.4. The correlation coefficient is remarkably low, 0.685, but looking at the graph obviously three cases (marked with a circle) disturbed the relation substantially.

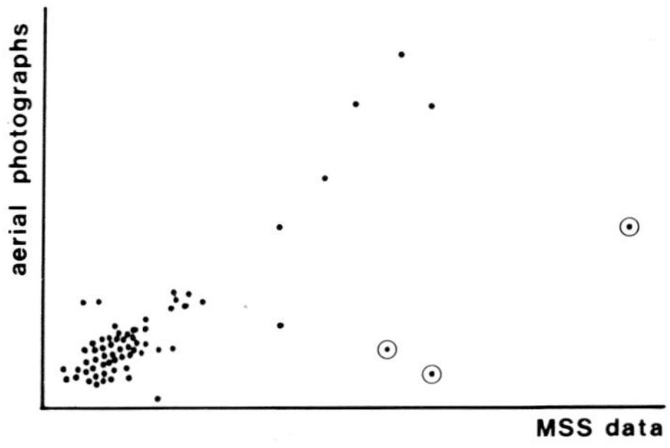

Figure 10.4 Relations between village size in Landsat data and aerial photographs.

In accordance with Section 7.5, these villages were flagged for manual inspection. And such a erroneous calculation would not occur for a manual demarcation of villages, which on the other hand, is an extremely slow process.

Forsaking these cases, the correlation coefficient is 0.918.

Regression equation

When establishing a mathematical relationship between number of inhabitants and size, more than 22 villages are needed. Hence, data collected under other field work and under other circumstances had to be utilized. Although this is not statistically suitable, it is more appropriate than no analysis at all, as long as too far-reaching inferences are not drawn.

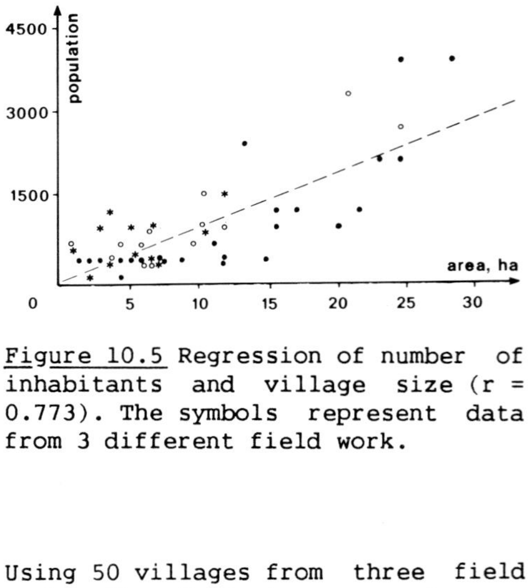

Figure 10.5 Regression of number of inhabitants and village size (r = 0.773). The symbols represent data from 3 different field work.

Using 50 villages from three field work, the following regression was established (Fig 10.5)

POPULATION = 93.2×AREA (ha)

The Durbin-Watson statistic is a test of the residual auto-correlations of the regression. The value given reacts if the errors are not uncorrelated, which is a preassumption when making the regression.

For 50 cases, the Durbin-Watson critical values, on a 0.05 level of significance, are 1.50 - 1.59. The present regression obtained the value 1.57, which implies that the data set is inconclusive for regression.

Also an attempt to adapt the data set to the logarithmic formula was carried out, but resulted in a considerably lower regression coefficient, and was not further utilized.

10.7 Relations between population and spectral density

IOD (integrated optical density) is defined as 'the sum of the pixel DGLs within a delineated object'. The spectral density is the IOD divided by the number of pixels, i.e. the 'mean DGL within the object'. Based only on the knowledge from Chapters 2 and 3, it is likely that total population in a village is better correlated to the spectral density than it is to size solely.

Iisaka & Hegedus (1982) assumed that the proportion of manufactured objects within an MSS pixel varies with the degree of urbanization, i.e. the density of the population. They also found a strong linear regression correlation between the population density and MSS 6 and MSS 7.

Table X.2 Correlation between number of households as measured from aerial photographs, size and IOD. The calculations are made based on figures from 33 villages in the study area.

	Number of households	
Size	0.910	(0.824-0.955)
IOD Mar4	0.911	(0.826-0.956)
IOD Mar5	0.914	(0.832-0.957)
IOD Mar6	0.917	(0.837-0.959)
IOD Mar7	0.916	(0.835-0.958)
IOD Mar7/Mar5	0.908	(0.820-0.954)

Considering the rather high correlation between size and households or population it is probably difficult to increase the accuracy by adding the spectral density to the size. Table X.2 denotes the correlation coefficients regarding size, IOD and population.

Although the correlation coefficient for the single band IODs at first seems to be higher than for size only, the difference is small and not significant.

It was impossible to establish a significantly better correlation between households and IOD, and hence the IOD approach will be discontinued in the present investigation.

10.8 Conclusions and discussion

Far below the original intentions of this study, the population data collected may be used as a demonstration of the methods, rather than for statistical determination of the accuracy of the presented methodology.

The correlation coefficient between settlement size in March 1979 (as automatically measured from Landsat data) and stated number of inhabitants in 1982 was 0.773. However, the correlation was also based on cases collected outside the study area, and from different field surveys.

The crowding index figures exhibited a large deviation, implying that some of the given answers were erroneous. The deviation of the growth rate figures cannot be interpreted as an indication of incorrect data. However, the magnitude of the growth rate is suspiciously high.

Although no proof could be presented, common sense propounds that inaccurate data should not, by coincidence, result in a high correlation. Therefore, it is not entirely improbable that the correlation between settlement size and population size in reality is higher than 0.773.

11 population maps

11.1 Background

As a final test of the proposed methods, they were applied to a census district outside the study area. A reason for choosing another area was that census data, on the district level, were needed. The division of census districts changed from the 1973 census to the 1983 census, and the only district left unchanged was the El Obeid rural council (Fig 11.1).

Having data from both dates, it was possible to project the population on 1979 and, thus, get a better idea of the possibilities. The major drawback of this approach was still the lack of accurate data enough on the village level.

Originally, this final test was planned to be carried out in one of the rural councils from which census data on lower levels could be obtained. Unfortunately, all the data that were possible to obtain were restricted to village councils, i.e. comprising two or more villages and of no value for establishing the size/population relationship.

As an inferior substitute the El Obeid rural council was used for the final test. This council is an immediate neighbour to the west of the study area. It comprises almost the same kind of vegetation and soils, except for some larger areas with non-cultivated savanna. The only physical difference is actually some small inselbergs rising some ten, maybe hundred, meters above the smoothly undulating Qoz.

During our field work in the area, which did not include any sufficient population investigations, we experienced no major differences in sizes, shapes and location of the settlements, compared with the study area. However, looking at the satellite data, the consistency of elliptic and circular forms noticed in the study areas, was not as evident in the El Obeid area. This might be a hint that the size/population re-

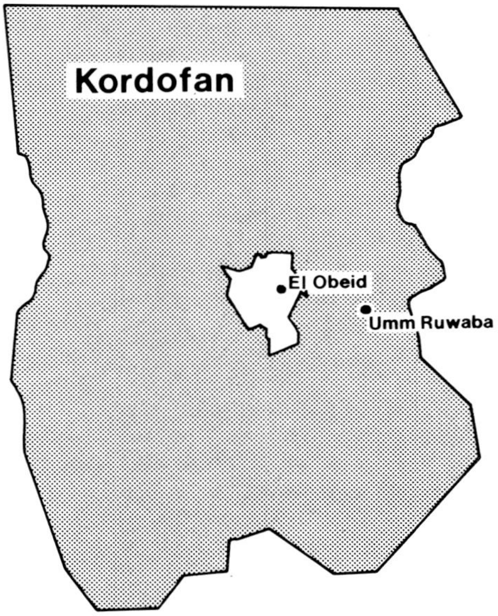

Figure 11.1 El Obeid rural council.

lation could be slightly different.

Table XI.1 denotes the available census data of the El Obeid town and rural council.

Table XI.1 Census data from the last two censuses of the El Obeid town and rural council (Department of Statistics 1980a and 1984).

	Town	Rural
settled pop 1973	90,073	88,233
settled pop 1983	139,446	100,012
Nomad pop 1973	-	4,973
Nomad pop 1983	2,082	9,594

From the figures of Table XI.1 the growth rates for El Obeid rural council might be calculated (Table XI.2). If the stated under-numeration of the census in 1973 is to taken into consideration also for the El Obeid area, a diffrerent rate is obtained.

Table XI.2 Growth rates from 1973 to 1983 according to Table IX.1. The 5% figures include an assumed under-enumeration 1973 of 5%.

	Original	incl 5 %
El Obeid town	4.47 %	3.96 %
El Obeid rural	1.26 %	0.78 %
El Obeid nomad	6.79 %	6.27 %
Total	3.11 %	2.61 %

From Table XI.1 and XI.2 the population in 1979 may be projected (Table XI.3).

Table XI.3 Population in El Obeid town and rural council 1979 as projected from the censuses in 1973 and 1979.

El Obeid town	119,394
Rural settled	97,068
Nomad	7,521
Total	225,803

The figures from the 1983 census provided also figures for households from which the crowding index could be calculated (Table XI.4). Unfortunately, these figures could not be projected to 1979, but they are of course much more stable than the crude population number.

11.2 Methods

The results of the methodology chapters implied a manual identification of the villages, followed by a computerized size determination.

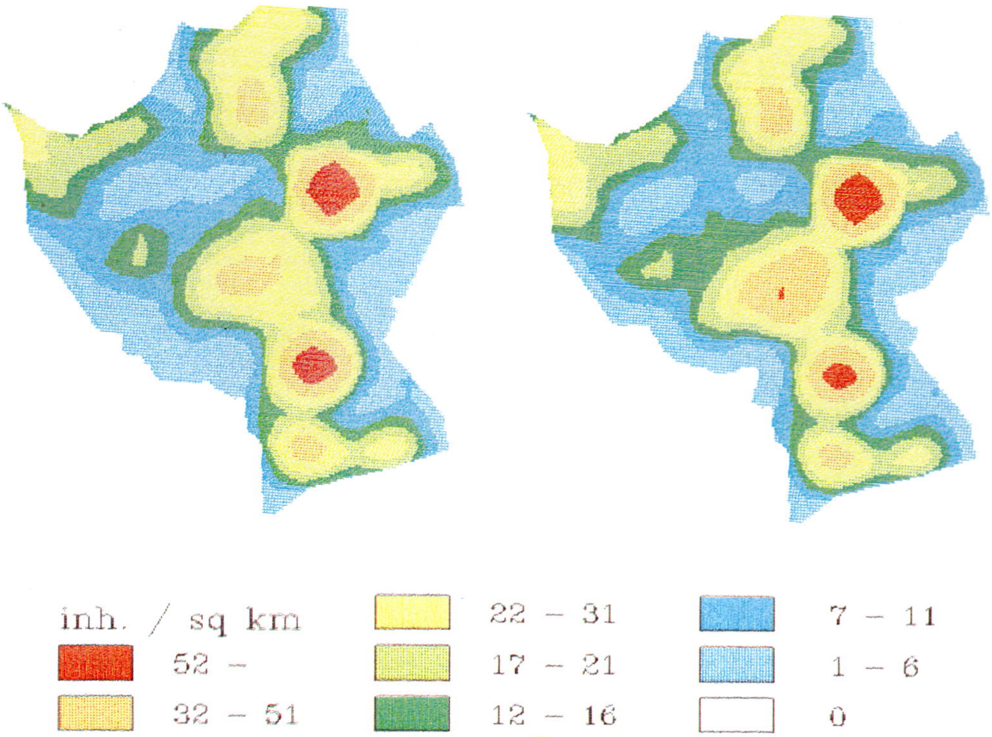

Figure 11.2 Population density 1972 (a) and 1979 (b) as calculated with remote sensing techniques.

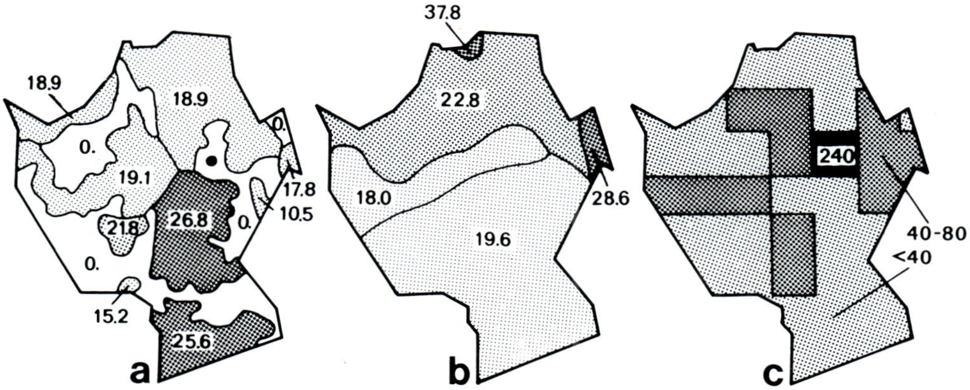

Figure 11.3 Population density a) 1956 (Doxiadis 1966), b) 1976 (Watson et al 1976) and c) 1973 (from Population Density map produced by Sudan Survey Department 1974).

Table XI.4 Crowding index for El Obeid town and rural council in 1983, figures calculated from the Department of Statistics (1984).

El Obeid town	6.10	persons per household
Rural settled	5.33	
Nomad	6.18	

Identification

In Chapter 8 it was pointed out that the properties of the principal components are strictly scene-dependent, and the favourable results were not possible to transfer to this area. The villages and their halos where quite different in the appearance, which resulted in difficulties when identifying the villages.

Hence, then identification was made in a slightly different manner than in the study area. The village were first identified in dry season data, where most villages were picked easily. Then, they were separated from inselbergs, which appeared blue in multi-temporal data.

Finally, they were also checked against the wet season data. This was a more circumstantial procedure than the one made in the original study, but probably with a similar fidelity.

It should also be noted that population mapping (identification + size measurements) consumed less than 8 man-hours of work for the entire El Obeid rural council (approx. 70,000 square km).

11.3 Maps

Figs 11.2a and 11.2b are the population density maps produced, and Figs 11.3a through 11.3c are other density maps, found in the literature.

Because of the unproven quality of Fig 11.2 it is impossible to thoroughly compare the content of the maps. It is also somewhat unfair to do so, since Fig 11.3a-c are not designed to provide the same geographical accuracy.

It should also be noted that the latter maps are based on data from different years, collected with different methods, and stratified in different ways. Nevertheless, it is interesting to imagine what could be done with the village population data in the digital form.

Village population numbers were placed in their correct pixel in the raster data. Then the raster was heavily smoothed with a low-pass filter. If the filter is non-boosting, and if the pixels are divided by the size of the pixel, a 'correct' population density value is given in every pixel.

The population density could be calculated based on area under cultivation, as Fig 11.3a, which is rather simple to map from remotely sensed data (see e.g. Helldén 1984). It could also be calculated based on strata, soil or vegetation (as Fig 11.3b).

The growth rate map (Fig 11.4) was produced by simply using the population density values in every pixel in the formula

$$\text{GROWTH RATE} = \sqrt[n]{\frac{\text{POP}_{\text{YEAR 1}}}{\text{POP}_{\text{YEAR 1+n}}}} - 1$$

11.4 Comparative statistics

Since there is no reliable village data for the mapped area, there is no real possibility to determine the accuracy for the maps, or to draw any far-reaching conclusions.

However, it may be interesting to just look at the figures obtained (Table XI.5), if calculating crude population numbers (based on a relationship with low fidelity, obtained

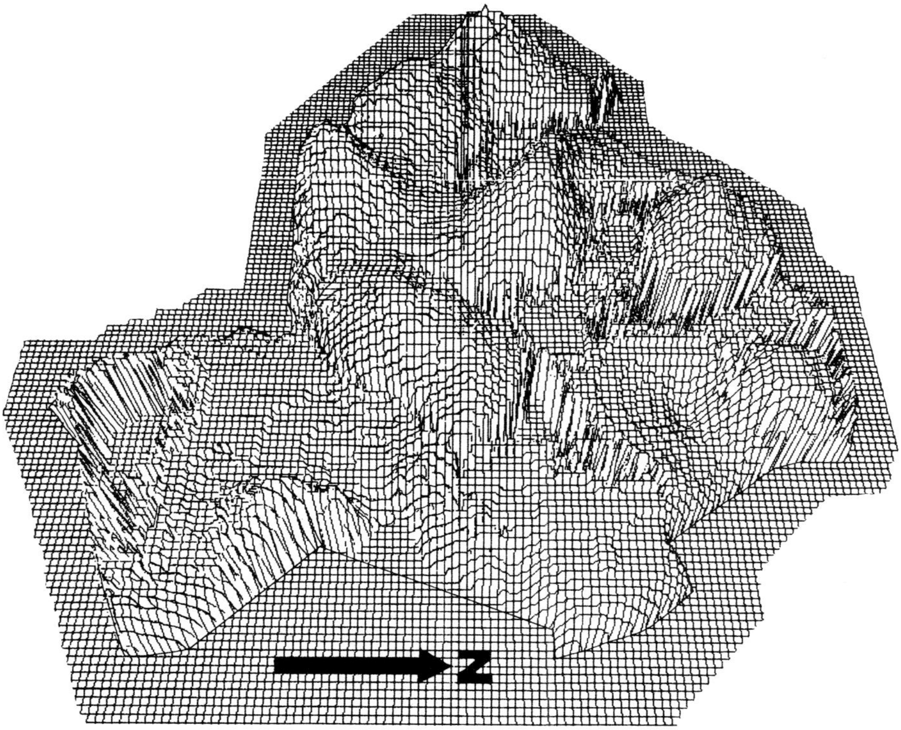

Figure 11.4 Growth rate 1972 - 1979 based on Fig 11.2. Extreme low values are approximately -14% and high values are +21%.

Table XI.5 Population and growth rate in El Obeid town and rural council 1979, based on censuses 1973 and 1983 (Department of Statistics 1980a and 1984), and the remote sensing technique presented.

	El Obeid	Rural settled	Rural nomad	Total rural	Total
Census pop 1973	94,577	92,647	5,222	97,869	192,446
RS pop 1973	-	100,059	-	100,059	194,636
Proj pop 1979	119,394	97,068	7,521	104,589	225,803
RS pop 1979	-	110,197	-	110,197	229,591
Census growth	3.96	0.78	6.27	1.11	2.70
RS growth	-	1.60	-	1.60	2.79

in another area), and compare them with projections based on two censuses 10 years apart (of which the first one was critized for its low accuracy). Hence, it might be by mere chance that some figures are close.

El Obeid town was not at all included in the size measurements, since it has a completely different structure than the rural settlements. Data regarding El Obeid is much easier to obtain than data for rural population, and could be done with other methods in a real application. Thus, the urban figures are denoted in Table XI.5. Also, total rural figures are denoted in the table, because most nomads have dwellings in some of the villages, and are therefore 'counted'.

11.5 Discussion

The interesting experiences faced in this final test of the population mapping, regards more the differences compared with the study area mapping, than the elusive similarity of obtain population figures.

The principal components were not outstanding as interpretation features, probably because data from different years were included in the K-L transform. For the same reason, the multi-temporal feature combination was not as good as in the study area.

It is probabyly adviseable to use multi-temporal data from the same year only, for manual interpretation of spectral and K-L transformed multi-temporal features.

Another experience was the minor change of village shape compared with the study area. After a short period of training it was not difficult to identify them, but it points to the necessity of closely investigate village appearance and, to use stratified size/population relations.

12 a look into the future

At the very end of the work with this thesis, an opportunity arose to test automatic village identification on a GOP-300 system at Context Vision, Linköping. Since the bulk of the study was already carried out, there was no possibility to utilize the obtained results for the population maps. Nevertheless, it was interesting to get a short glance at the image processors of the future.

12.1 GOP

The population study was made with 'classical' remote sensing methods, or with small deviation from the 'classics'. GOP might be regarded as a part of the development in the field of Artificial Intelligence, although it is not a complete expert system for remote sensing.

The hard-ware benefits are not at all considered here. The soft-ware is based on a concept fairly different from conventional remote sensing systems, but cannot be described in detail.

Concept

The main advantage is the theoretical base of the filtering technique, and the design of the kernel (window coefficients), or rather the set of kernels. The filtering output is not one, but two new features. They have a polar (complex) representation, where

$$y = \underline{z} \cdot e^{(i\psi)}$$

Thus, the pixels in the output images are

in image $\underline{z} = |y|$ (real part)

in image $\psi = \arg y$ (imaginary part)

For example, when computing the local orientation (Cf filtering in e.g. Schowengerdt 1983) the two components are

1) the argument, $\underline{z}$, which is the dominating orientation within the window;

2) the magnitude, ψ, which is the accuracy with which the orientation is determined.

The argument, either it describes the local orientation or e.g. local variance, may be represented by the angle of a vector. When the vector denotes the variance, the variance must be transformed from its original ratio scale to a polar scale.

The magnitude is the length of the same vector. Hence, every pixel in the output image is a vector with a certain orientation and length.

When the images are displayed on the monitor, the argument angle is displayed as the respective colour of the colour circle, and the magnitude is the intensity of the colour.

This concept has a row of advantages, of which only a few are utilized in the operators used here. For a more complete description, see e.g. Knutsson & Granlund (1983), Knutsson et al (1983), Hedlund et al (1982) or Context Vision (1985).

Operators used

Since every pixel is a vector, its angle represents a variable value, e.g. orientation, variance or frequency. This variable can be illustrated in a circular diagram. Each operator adopted in the present test is depicted in such a diagram.

- local orientation; is simply the dominating orientation within the window. However, it has a different representation in the diagram (Fig 12.1), since 'west' and 'east' is the same thing, namely a horizontal orientation.
- rotation consistency; is based on the output vectors of the local orientation. It is sensitive to different orientation patterns (Fig 12.2).
- divergence consistency; also works on the orientation vectors, but reacts on another set of patterns (Fig 12.3).

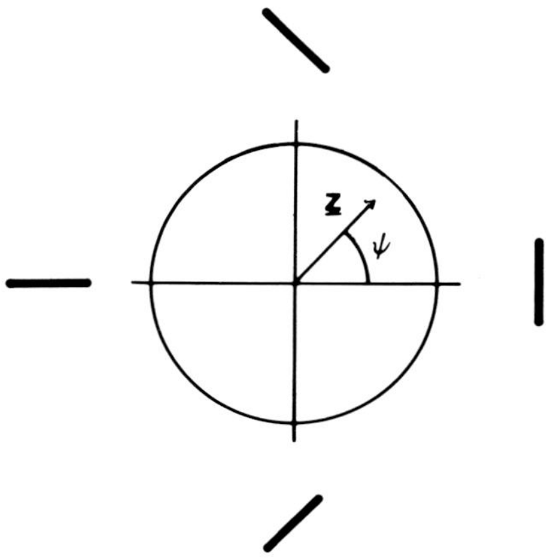

Figure 12.1 Relation between output vectors and local orientation.

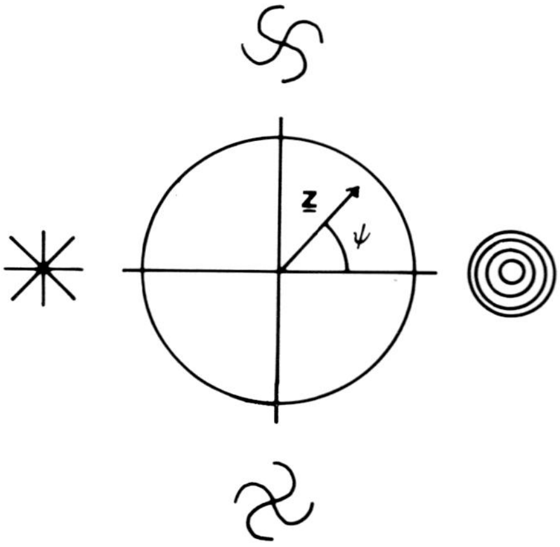

Figure 12.2 Relation between output vectors and patterns enhanced by the rotation consistency operator.

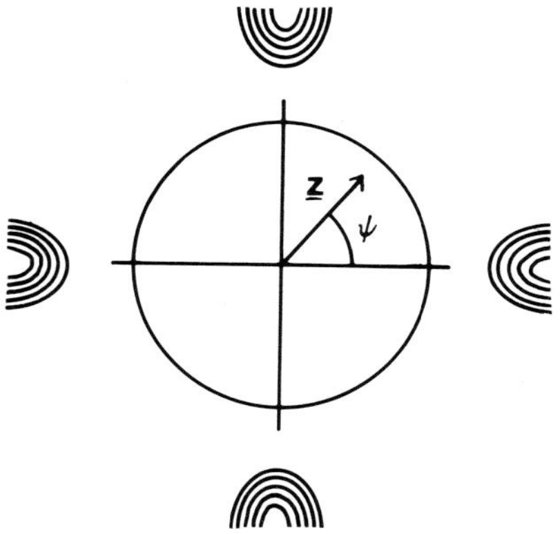

Figure 12.3 Relation between output vectors and patterns enhanced by the divergence consistency operator.

- phase; which works on the original image. Considering the DGL variations in different directions as the frequencies of a signal, it calculates in which phase of a period, that dominates the window. In reality, it detects dark lines, light lines and edges. Since dark-light edges and light-dark edges are regarded equal, the output vectors only utilizes a semicircle in the diagram (Fig 12.4).

- 'feature 1';* operates on the original images, and is sensitive to different dot patterns (Fig 12.5).

Except for the operators described and hybrids of them, there are several more operators. A large group of them embraces the dominating frequency, which is closely related to inter-pixel variability, one of the texture features in Chapter 6.

But, since features involving village internal texture already have been examined, the GOP test was concentrated on the texture, or context, created by the village as an object. This implies attempts to find villages as dark spots in a light background (village halo).

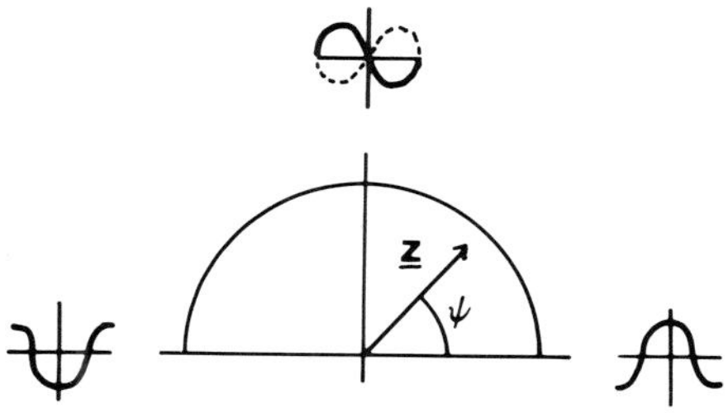

Figure 12.4 Relation between output vectors and the phase in the period for the dominating frequency.

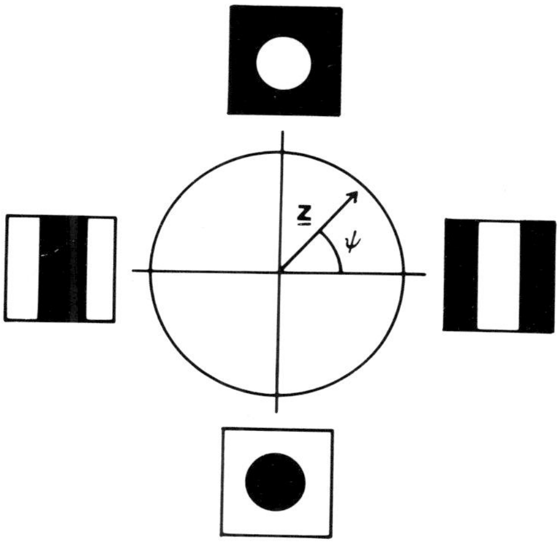

Figure 12.5 Relation between output vectors and patterns enhanced by the 'feature 1' operator.

The effect of some transformations (made on Mar6), can be seen in Figs 12.6 through 12.10. Since the GOP images are originally in a polar representation, they had to be transformed into Cartesian representation, to fit the M-L classification. Hence, they do not have the same appearance in the figures as they originally had in the GOP-300 system.

---

*Feature 1 is an intermediate name, that will be change when the operator is released. Do not confuse with the general feature concept.

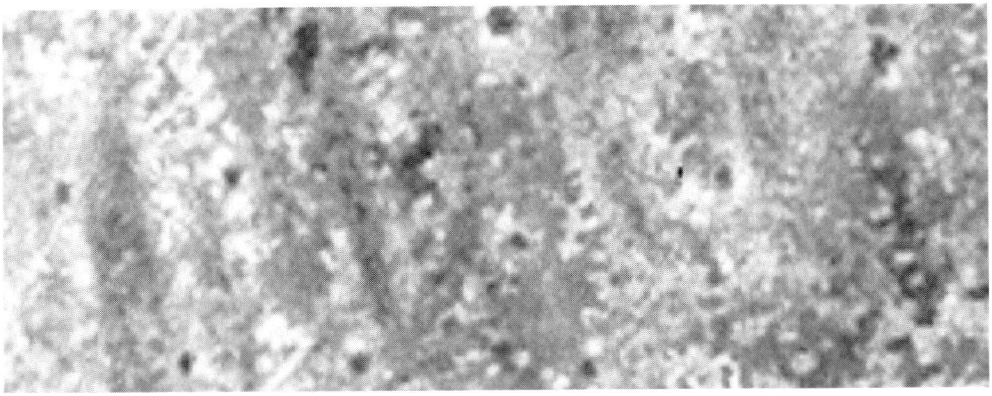

Figure 12.6 Original sub-area (Mar6).

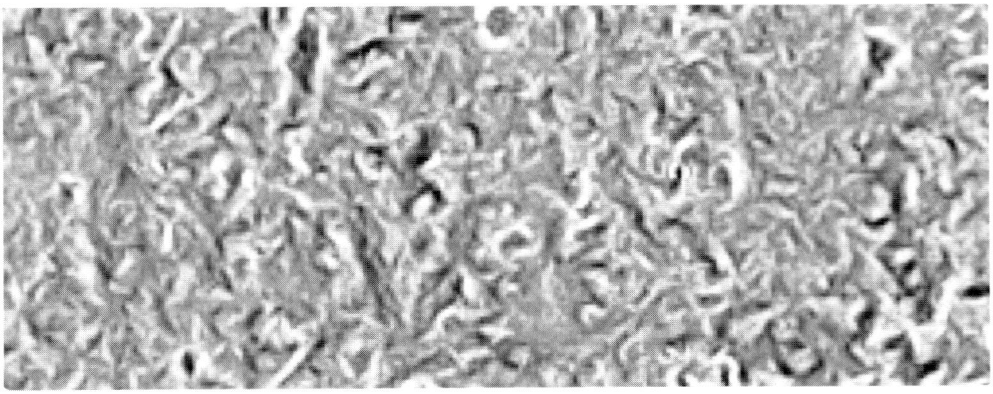

Figure 12.7 Real part of the complex image produced by the phase operator applied to the original image.

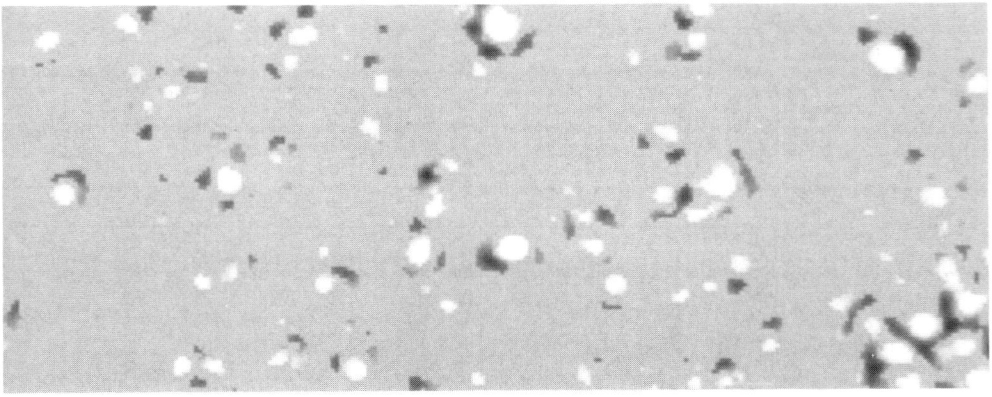

Figure 12.8 Imaginary part of the complex image produced by the rotation consistency operator applied to a 'feature 1' image.

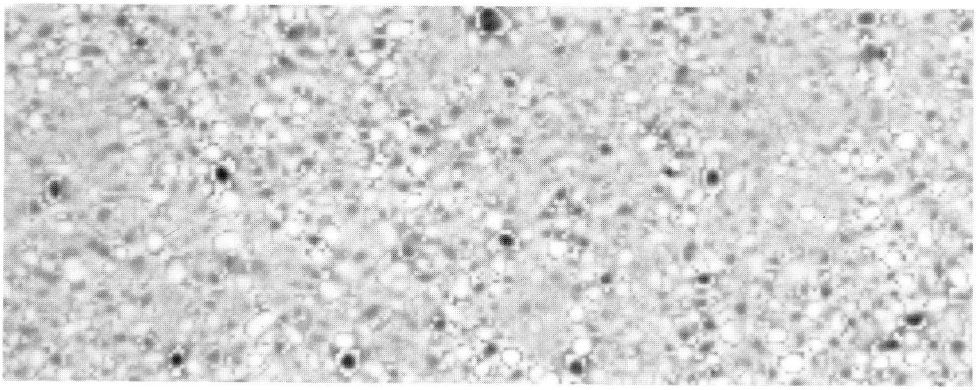

Figure 12.9 Imaginary part of the complex image produced by the 'Feature 1' operator applied to the original image.

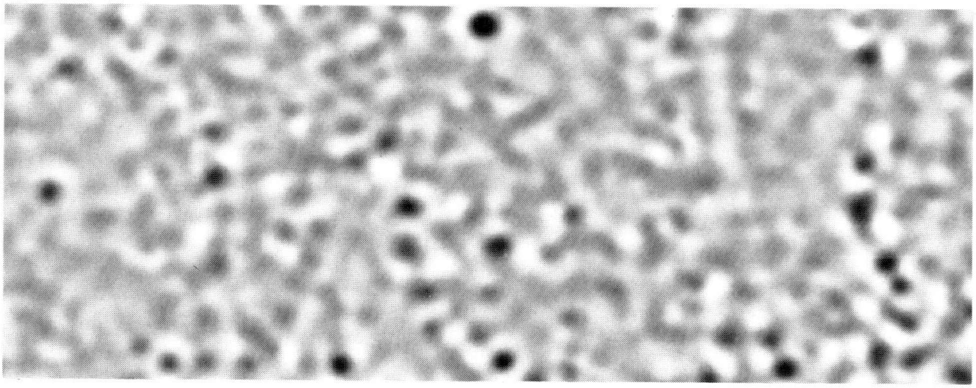

Figure 12.10 Real part of the complex image produced by the rotation consistency operator applied to a local orientation image.

12.2 Classification

These four images were, together with four spectral bands, exposed to M-L classifications, in accordance with previous classifications (Table XII.1).

The consistency features had identification percentages and over-classification similar to the best combined automatic methods (Section 8.2). But the real advantages arose when combining 'feature 1' with the spectral bands. Keeping the identification rate, the over-classification dramatically dropped.

Table XII.1 Result of the most powerful M-L classifications involving spectral bands and some transformations with GOP operators. Cf Section 8.2.

Image	Identified villages	% of area
Spectral bands	100 %	15.7
Phase	100 %	20.7
'Feature 1' + rotation	96 %	15.6
'Feature 1'	96 %	11.6
Orientation + rotation	96 %	8.6
Spectral + 'Feature 1'	100 %	3.2

But for a more complete comparison, also a 'quick-and-dirty' erosion-dilation was performed, followed by a segmentation. Thus, it is possible to compare the obtained AND-OR values with the conclusions in 8.5 (Fig 12.12).

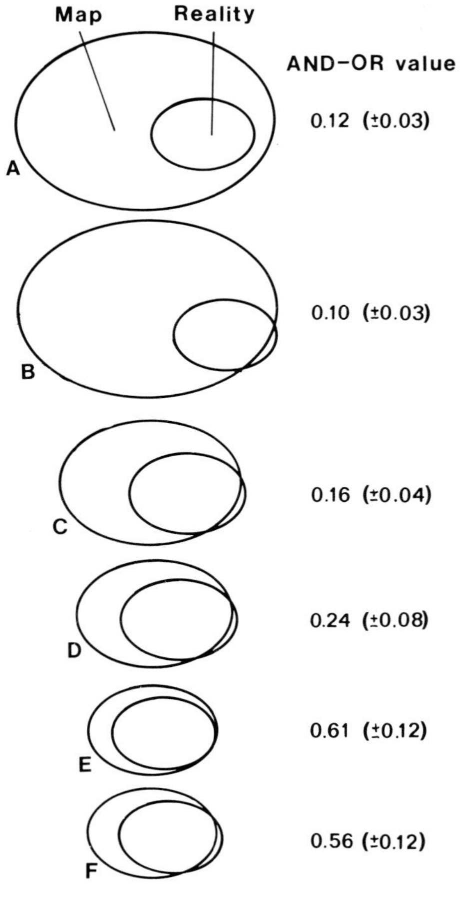

Figure 12.11 AND-OR values for classification of the image in a) Fig 12.7, b) 12.8, c) 12.9, d) 12.10, e) 12.10 + four spectral bands and f) same as e) but also exposed to morphometric discrimination.

12.3 Conclusions

Involving GOP transformations in the M-L classifications considerably decreased the over-classification, and increased the accuracy. The morphometrics did the classification a disservice by removing some of the identified villages. This is probably caused by the fact that the 'feature 1' is a stronger shape measure than the adopted morphometrics.

Probably, even higher AND-OR values could be obtained with a more carefully performed pre-processing and analyses, prior to the classification. However, it is questionable whether the methods would be competitive to the manual interpretation results. Nevertheless, the operators comprised in GOP are, for the present application, superior to 'classical' remote sensing methods, which makes the system highly interesting.

13 synthesis

The final chapter of this study comprises the conclusions and discussions about the usefulness of the proposed methods, and also speculates briefly about future approaches and research trends.

13.1 Methodology

The estimation of population in a semi-arid area in the Sudan from Landsat MSS data, involves two major efforts, to identify the human settlements and to relate the settlement sizes to population sizes.

A major part of the study is the comparison between a variety of automatic village identification methods and manual village interpretation.

The tests of automatic methods signified that multi-temporal feature combinations outruled single-season features for village detection, and the use of arithmetical and textural features did not significantly add any discriminatory power.

Post-classification shape measurements of classified objects increased the mapping accuracy, but the manual interpretation of villages was strongly superior to the automatic methods.

Due to the lack of sufficient 'ground truth', the strength of the relationship between settlement size and population size could not be verified.

TM and Spot data

The increased geometric resolution of TM and Spot data is not expected to enhance the village classification accuracy substantially, since this is not mainly dependent on settlement size. On the other hand, the increased radiometric and spectral resolution might be helpful in this respect.

The geometric resolution might assist the morphometric discrimination of villages, since the settlement shapes are better preserved.

13.2 Tradition and trends in image processing

Much of the presented discussion regards a dichotomy between manual and computer-assisted methods, which is based on the design of today's image processing systems. However, this dichotomy is not a desired one. Estes et al (1983) declare:

> 'A clear need now exists to critically re-examine the basic elements and principles of image analysis that are common to both manual- and computer-assisted techniques. Indeed, due to individuals in the diverse group of disciplines involved, which includes the application-oriented disciplines such as, forestry, range management, hydrology, geology, and geography on the one hand and electrical engineering and computer science on the other, artificial barriers between manual and computer-assisted techniques have been erected. These barriers are false and have only made it more difficult to discern either fundamental similarities or differences between the two approaches.
>
> The objective of this discussion is to provide a broad conceptual framework, a paradigm wherein the basic linkages between what has traditionally been referred to as manual or human image analysis and computer-assisted or automated image are put into perspective'.

One implication of the mentioned barriers is that none of the image processors used in the present study embodied convenient functions for support of manual interpretation. The only manual identification anticipated by the processing systems was training area definition as input for automatic classification.

Hints

The manual interpretation of satellite images cannot entirely return to analog techniques. It is essential that the advantages of the digital information is utilized.

But, along with the development of expert systems and advanced algorithms for pattern recognition, the manufacturers of image proces- sing systems should develop and soft-ware for semi-automatic inter- pretation. It may be functions were the operator is supported by e.g. advanced textural and contextual algorithms when delineating obscure land cover classes, or a tasteful post-classification editor to manually surpress mis-classified areas. Or it might be an intelligent inter-action between supervised and unsupervised classification.

The hard-ware manufacturer should develop e.g. a large monitor that <u>is</u> the very table-top itself and suitable for manual interpretation, and not a user-hostile monitor just sitting on the work station table.

13.3 Census from heaven?

The second aim of the study regards the utilization of the proposed remote sensing techniques for population mapping. One can foresee a variety of uses of the maps,

- for intercensal counts;
- for census support;
- for geographical population modelling.

Unfortunately, it was not possible to precisely determine the sensitivity of the instrument. But, it is self-evident that the accuracy of the methods will vary with the objectives, the data used and the area in which the methods are employed. Hence, it is necessary to investigate the fidelity from time to time.

Intercensal counts

Intercensal estimates, made on a sampling basis, are mainly carried out because of the census limitations. Censuses are often gigantic operations performed at long intervals, and the intercensal counts are needed to support the population projection from the latest census.

For such traditional estimates, remote sensing techniques may be utilized for determination of sampling frames. For a discussion about sampling frames, see e.g. Hanuschak & Morrissey (1977).

Using the proposed methodology, intercensal counts may instead be performed as total surveys with rough inhabitant numbers for each settlement. It may also be conducted at a comparatively low cost.

Another attractive application related to intercensal counts, is the monitoring of drought impact on the human population.

During the rising drought in northern Kordofan 1983/84 directors from the Ministry of Agriculture and Irrigation, together with FAO/WFP, estimated the immediate food aid needed for resident and displaced population. This was done through survey flights and field trips, where the vegetation cover, etc was collated (Darag 1984).

Similarly, remote sensing techniques are possible to use not only to monitor vegetation cover and prewarn of drought impact and famine, but also to estimate the number of people struck and to monitor gross refugee movements.

However, still more research is needed on such aspects, and also the possibility to monitor abandonment of settlement, which is an essential indicator of desertification and/or severe drought.

It is also interesting to note that the used methods are tested in an area typical of the Sahel zone, which is the zone mostly struck by the latest well-known drought and famine catastrophes. And, the entire Sahel zone is similarly constructed regarding livelihood systems, settlement patterns, vegetation and soils. Hence, it is plausible that the proposed methods are applicable to the entire zone.

Census

The main use of the suggested methods for traditional censuses is to provide census maps. The omission of settlements in census procedure is one of the main disturbances of the census accuracy, especially in countries like the Sudan, where the base maps are extremely poor. But, as a subsitute for traditional censuses the methodology is insufficient, since it only considers gross population or population density, and not any of the other demographic parameters.

Geographical modelling

The geographical accuracy of the population density described in Chapter 11, together with digital representation of the data, is the major efforts of the presented methods. As seen clearly in Olsson (1985), precise geographic information about population pressure may form the most basic information layer when modelling desertification.

references

Abrams M.J. 1980: Lithologic mapping. In Siegal & Gillespie (eds): Remote sensing in geology. John Wiley & Sons, New York, 702 pp.

Allen J.A. and Alemayehu T. 1975: Rural population estimates from air photographs: An example from Wolamo, Ethiopia. ITC Journal, no 1, pp 85-100.

Alm G. 1985: Land use classification of aerial photographs using texture and spectral information. Proceedings of the 4th Scandinavian Conference on Image Analysis, Trondheim, Norway, pp 373-382.

Anderson D.E. and Anderson P.N. 1973: Population Estimates by humans and machines. Photogrammetric Engineering 39, pp 147-154.

Austin R.F. 1984: Measuring and comparing two-dimensional shapes. In Gaile & Wilmott (eds): Spatial statistics and models. D Reidel Publishing Company, 482 pp.

Babiker A.B. 1982: Urbanization and desertification in the Sudan, with special reference to Khartoum. GeoJournal 6.1, pp 69-76.

Barbour K.M. 1961: The republic of the Sudan, a regional geography. University of London Press Ltd, 292 pp.

Bengtsson E., Eriksson O., Holmquist J., Nordin B. and Stenkvist B. 1979: High resolution segmentation of cervial cells. Journal of Histochemistry and Cytochemistry 27, pp 621-628.

Bondestam L. 1973: Some notes on African statistics. Scandinavian Inst of African Studies, rep 18, 50 pp.

Born M. 1965: Zentralkordofan, Bauern und Nomaden in Savannengebieten des Sudan. Marburger Geographischen Schriften, Heft 25, 252 pp.

Boyce R.R. and Clark W.A.V. 1964: The concept of shape in geography.

Geographical Review 54, pp 561-572.

Bressanin G. and Erickson J. 1973: Data processing systems for earth resources survey. ESRO CR-295, European Space Research Organisation, Neuilly, France.

Brugioni D.A. 1983: The census: It can be done more accurately with space-age technology. Photogrammetric Engineering and Remote Sensing, no 9, pp 1337-1339.

Bunge W. 1962: Theoretical geography. Lund Studies in Geography, ser C, no 1, University of Lund, 210 pp.

Bushra el- S. 1967: Factors affecting settlement distribution in the Sudan. Geografiska Annaler 49 (ser. B), Lund, pp 10-24.

Castlemann K.R. 1979: Digital image processing. Prentice-Hall, Englewood Cliffs, 429 pp.

Clayton C. and Estes J.E. 1980: Image analysis as a check on census enumeration accuracy. Photogrammetric Engineering and Remote Sensing 46, pp 757-764.

Clarke K.C. 1984: Shape-assisted land-use classification for remotely sensed data. Proceedings of 18th Int Symp on Remote Sensing of Environment, Paris, France, October 1-5, pp 1517-1525.

Colvocoresses, A.P. 1977: Proposed parameters for an operational Landsat. Photogrammetric Engineering 43, pp 1139-1145.

Colwell R.N. (ed) 1983: Manual of remote sensing. American Society of Photogrammetry, Fall Church, 2440 pp.

Context Vision 1985: Advanced image processing - practice sessions on the GOP-300 system. Context Vision, Linkoping.

Curran P.J. 1985: Principles of remote sensing. Longman, London, 282 pp.

Darag A. 1984: Drought, desertification and desert encroachment in Kordofan and Darfur regions. Ministry of Agriculture and Irrigation, Khartoum.

Davies W.A. and Peet F.G. 1977: A method of smoothing digital thematic maps. Remote Sensing of Environment 6, pp 45-49.

Dayal H.H. and Khairzada B.A. 1976: The first national demographic survey of Afghanistan: the role played by air photos and photo-counting techniques. ITC Journal, no 1, pp 84-97.

DECARP 1976: Sudan's desert enchroachment control and rehabilitation programme. The General Administration for Natural Resources, Ministry of Agriculture, Food and Natural Resources, and the Agricultural Research Council, National Council for Research, in collaboration with UNEP, UNDP and FAO, 227 pp.

Demeny P. 1968: The demography of the Sudan: An analysis of the 1955/56 census. In Brass (ed): The demography of Tropical Africa. Princeton University Press, Princeton, 540 pp.

Department of Statistics 1980a: Second Population Census 1973, Volume II, Population dynamics. Khartoum, 260 pp.

Department of Statistics 1980b: Population Statistics, Kordofan. Khartoum, 121 pp.

Department of Statistics 1984: Third Population Census (Unpublished).

Doxiadis 1966: Land and water use survey in Kordofan Provinces of the Republic of the Sudan. Doxiadis Associates, Athens, Greece, DOX-SUD-A48, 92 pp.

Dregne H.E. 1976: Soils of arid regions. Elsevier, Amsterdam, 230 pp.

Dye R.H. 1975: Restoration of Landsat images by discrete two-dimensional deconvolution. Proceedings of the 10th Int. Symp. on Remote Sensing of the Environment, Ann Arbor, pp 725-730.

Estes J.E., Hajic E.J. and Tinney L.R. (eds) 1983: Fundamentals of image analysis: Analysis of visible and thermal infrared data. In Colwell (ed): Manual of Remote Sensing. American Society of Photogrammetry, Falls Church, 2440 pp.

Eyre L.A., Adolphus B. and Amiel M. 1970: Census analysis and population studies. Photogrammetric Engineering 36, pp 460-466.

Foresti C. and de Mendonca F. 1974: Demographic inference using ERTS images. Sem. Space Appl. Dir. Interest to Develop. Countries, COSPAR, Sao Jose dos Campos, Brasil, vol 2, pp 239-255.

Forster B.C. 1980: Urban residential ground cover using Landsat digital data. Photogrammetric Engineering and Remote Sensing 46, pp 547-558.

Forster B.C. 1982: The derivation of approximate equations to correct for Landsat MSS point spread function. Proceedings of the ISPRS Symp, Commission I, Primary data aquisition, Australia, pp 6-10.

Forster B.C., de Kantzow D.R. and McMahon G. 1984: Urban and heterogeneous environments workshop. Proceedings from 3rd Australasian Remote Sensing Conference, Queensland, pp 765-768.

Forster B.C. and Trinder J.C. 1984: An examination of the effects of resampling on classification accuracy. Proceedings from 3rd Australasian Remote Sensing Conference, Queensland, pp 106-115.

Frank T.D. 1984: The effect of change in vegetation cover and erosion patterns on albedo and texture of Landsat images in a semiarid environment. Annals of the Association of American Geographers, vol 74, no 3, pp 393-407.

Fu K.S. 1976: Pattern recognistion in remote sensing of the Earth's resources. In Bernstein (ed): Digital image processing for remote sensing. IEEE Press, New York, 473 pp.

van Genderen J.L. and Lock B.F. 1977: Testing land use map accuracy. Photogrammetric Engineering and Remote Sensing, vol 43, pp 1135-1137.

Gibril M.A. 1979: Evaluating census response errors - A case study for the Gambia. OECD, Paris, 98 pp.

Goodenough D.G. and Goldberg M. 1979: Processing of remote sensing data. Canadian Centre for Remote Sensing, 130 pp.

Gurney C.M. 1982: The use of contextual information to detect cumulus clouds and cloud shadows in Landsat data. International Journal of Remote Sensing 3, no 1, pp 379-388.

Gurney C.M. and Townshend J.R.G. 1983: The use of contextual information in the classification of remotely sensed data. Photogrammetric Engineering and Remote Sensing 49, no 1, pp 55-64.

Haack B.N. 1983: An analysis of Thematic Mapper simulator data for urban environments. Remote Sensing of Environment 13, pp 265-275.

Hanuschak G.A. and Morrissey K.M. 1977: Pilot study of the potential contributions of Landsat data in the construction of area sampling frames. Research and Development Branch, Research Division, Statistical Reporting, USDA, Washington D.C., 66 pp.

Hammer Digernes T. 1979: Wood for fuel - Energy crisis implying desertification. Major Thesis in Geography, University of Bergen, 124 pp.

Haralick R.M. and Fu K.S. 1983: Pattern recognition and classification. In Colwell (ed): Manual of Remote Sensing. American Society of Photogrammetry, Falls Church, 2440 pp.

Haralick R.M. and Shanmugam K.S. 1974: Combined spectral and spatial processing of ERTS imagery data. Remote Sensing of Environment 3, pp 3-13.

Harris R. 1985: SIR-A imagery of Tunisia and its potential for population estimation. International Journal of Remote Sensing 6, no 7, pp 975-978.

Hedlund M., Granlund G.H. and Knutsson H. 1982: A consistency operation for line and curve enhancement. Proc of IEEE computer Society Conference on Pattern Recognition and Image Processing, Las Vegas.

Helldén U. 1978: Evaluation of Landsat-2 imagery for desertification studies in Northern Kordofan, Sudan. LUNI, Rapporter och Notiser 38, 57 pp.

Helldén U. 1979: En test av Landsat-1 bild och digitala data för kartering av vegetation och markanvändning i Skåne. LUNI, Rapporter och Notiser 44.

Helldén U. 1980: A test of Landsat-2 imagery and digital data for thematic mapping illustrated by an environmental study in Northern Kenya. LUNI, Rapporter och Notiser 47, 63 pp.

Helldén U. 1981: Satellite data for regional studies of desertification and its control, approaches to to rehabilitation of degraded ecosystems in Africa. LUNI, Rapporter och Notiser 50, 20 pp.

Helldén U. 1984: Drought impact monitoring - A remote sensing study of desertification in Kordofan, Sudan. LUNI, Rapporter och Notiser 61, 61 pp.

Henin R.A. 1968: Second thought on Sudan's population census. In Caldwell & Okonjo (eds): The population of Tropical Africa. Columbia University Press, New York, 457 pp.

Holben B.N. and Justice C.O. 1980: An examination of spectral band ratioing to reduce the topographic effect on remotely sensed data. NASA/GSFC Tech Mem 80640.

Hord R.M. 1982: Digital image processing of remotely sensed data. Academic Press, New York.

Holz R.K., Huff D.L. and Mayfield R.C. 1969: Urban spatial structure based on remote sensing imagery. Proceedings of the 6th Int. Symp. on Remote Sensing of the Environment, Ann Arbor, Michigan, pp 819-830.

Hsu S. 1978: Texture-tone analysis for automated land-use mapping. Photogrammetric Engineering and Remote Sensing 44, no 11, pp 1393-1404.

Hsu S. and Burright R.G. 1980: Texture perception and the RADC/Hsu texture feature extractor. Photogrammetric Engineering and Remote Sensing 46, no 11, pp 1051-1058.

Hubel D.H. and Wiesel T.N. 1979: Brain mechanisms of vision. Scientific American 241, pp 150-162.

Hyde R.F. and Vesper N.J. 1983: Some effects of resolution cell size on image quality. Landsat Data User's Note 29, pp 9-12.

Ibrahim F.N. 1978: Anthropogenic causes of desertification in Western Sudan. GeoJournal 2.3, pp 243-254.

IBRD 1975: Review of the statistical system of the Democratic Republic of the Sudan. International Bank for Reconstruction and Development, Report no 535-SU.

Iisaka J. and Hegedus E. 1982: Population Estimation from Landsat imagery. Remote Sensing of Environment 12, pp 259-272.

Irons J.R. and Petersen G.W. 1981: Texture transforms of remote sensing data. Remote Sensing of Environment 11, pp 359-370.

Jaakkola S., Rosenholm D. and Isaksson L. 1984: SPOT-simulation - An evaluation for forest inventory and mapping purposes. Proceedings of the 18th Int. Symp. on Remote Sensing of the Environment, Paris, France, pp 521-528.

Jensen J.R. 1979: Spectral and textural features to classify elusive land cover at the urban fringe. Professional Geographer 31, no 4, pp 400-409.

Jensen J.R. (ed) 1983: Urban/suburban land use analysis. In Colwell (ed): Manual of Remote Sensing. American Society of Photogrammetry, Falls Church, 2440 pp.

Jensen J.R. and Toll D.L. 1982: Detecting residential land-use development at the urban fringe. Photogrammetric Engineering and Remote Sensing 48, no 4, pp 629-643.

Kauth R.J., Pentland A.P. and Thomas G.S. 1977: BLOB: An unsupervised clustering approach to spatial preprocessing of MSS imagery. Proceedings of the 11th Int. Symp. on Remote Sensing of the Environment, Ann Arbor, pp 1309-1317.

Kauth R.J. and Thomas G.S. 1976: The tasseled cap - Graphic description of the spectral-temporal development of agricultural crops as seen by Landsat. Proceedings of the Symposium on Machine Processing of Remotely Sensed Data, LARS, Purdue University.

Knutsson H. and Granlund G.H. 1983: Texture analysis using two-dimensional quadrature filters. IEEE Computer Society Workshop on Computer Architechture for Pattern Analysis and Image Database Management, pp 206-213.

Knutsson H., Wilson R. and Granlund G. 1983: Anisotropic nonstationary image estimation and its applications: Part I - restoration of noisy images. IEEE transactions on communications, vol COM-31, no 3, pp 388-397.

Kpedekpo G.M.K. 1982: Essentials of demographic analysis for Africa. Heinemann, London, 210 pp.

Kraus S.P., Senger L.W, and Ryerson J.M. 1974: Estimating population from photographically determined residential land use types. Remote Sensing of Environment 3, pp 35-42.

Langrebe D. 1971: System approach to use of remote sensing. Int Workshop on Earth Resources Survey Systems. NASA SP-283, pp 139-154.

Lindell T. 1983: Microcomputerized image processing of satellite data for water quality purposes. Proceedings of the 17th Int. Symposium on Remote Sensing of Environment, Ann Arbor, Michigan.

Lindell T. 1985: Metod for vattenkvalitetsbedömning från satellitdata. SNV PM 1891, Uppsala.

Lo C.P. and Chan H.F. 1980: Rural population estimation from aerial photographs. Photogrammetric Engineering and Remote Sensing 46, pp 337-345.

Lunden B. 1977: Mapping of surficial deposits from air photographs. SGU, ser C, no 738, 100 pp.

Monsted M. and Walji P. 1978: A demographic analysis of East Africa - A sociological interpretation. Scandinavian Institute of African Studies, Uppsala, 211 pp.

Morgan R.W. 1984a: Further comments on "The Census". Photogrammetric Engineering and Remote Sensing 50, no 1, pp 80-81.

Morgan R.W. 1984b: The census. Photogrammetric Engineering and Remote Sensing 50, no 11, pp 1609-1610.

Morrison A. 1980: Landsat digital

data for population estimation: testing the technique with grid-square census data in the Scottish Highlands. Proceedings of the 14th Int. Symp. on Remote Sensing of the Environment, Costa Rica, pp 1745-1749.

Morrow-Jones H.A. and Watkins J.F. 1984: Remote Sensing technology and the U.S. census. Photogrammetric Engineering and Remote Sensing 50, no 2, pp 229-232.

NAS 1977: Resource sensing from space, prospect for developing countries. National Academy of Science, Washington D.C., 201 pp.

Nordbeck S. 1971: Urban allometric growth. Geografiska Annaler 53 B, pp 54-67.

Novikoff G. 1983: Desertification by overgrazing. Ambio 2, no 2, pp 102-105.

Ogrosky C.E. 1975: Population estimates from satellite imagery. Photogrammetric Engineering and Remote Sensing 41, pp 707-712.

Olsson K. 1985: Fuelwood demand and supply in the Umm Ruwaba/Er Rahad region in N Kordofan, the Sudan. LUNI, Rapporter och Notiser 64, 74 pp.

Olsson L. 1985: An integrated study of desertification. Meddelande från Lunds Universitets Geografiska Institution, Avh XCVIII, 200 pp.

Olsson L. and Stern M. 1981: Large area data sampling, for remote sensing applications and statistical analysis of environment. LUNI, Rapporter och Notiser 49, 70 pp.

Olorunfemi J.F. 1983: Towards a philosophy of population census in Nigeria: remote sensing inputs. International Journal of Remote Sensing 4, pp 731-738.

Olurunfemi J.F. 1984: Land use and population: A linking model. Photogrammetric Engineering and Remote Sensing 50,' no 2, pp 221-227.

Paul C.K. 1984: More on the census. Photogrammetric Engineering and Remote Sensing 50, pp 1610-1612.

Pratt W.K. 1978: Digital Image Processing. Wiley, New York.

Rapp A. and Helldén U. 1979: Research on environmental monitoring methods for land use planning in African drylands. LUNI, Rapporter och Notiser 42, 124 pp.

Reeves E.B. and Frankenberger T. 1981: Farming systems research in North Kordofan, Sudan. University of Kentucky, Lexington, 72 pp.

Reeves R.G., Anson A. and Landon D. (eds) 1975: Manual of remote sensing. The American Society of Photogrammetry, Fall Church, 2123 pp.

Reining P. 1974: Human settlement patterns in relation to resources of less developed countries. Proceedings of Seminar on Space Applications of Direct Interest to Developing Countries, San Jose dos Campos, Brazil, pp 214-238.

Reining P. (ed) 1978: Handbook of desertification indicators. A.A.A.S., Washington D.C., 141 pp.

Reining P. 1980: Challenging desertification in West Africa. Papers in International Studies, Africa series no 39, Ohio University, 165 pp.

Reining P. and Egbert D. 1975: Interactive multi-spectral analysis of more than one Sonrai village in Niger, West Africa. Earth Resources Survey Symposium, Houston, NASA TMX-56168, pp 1811-1821.

Rosenfeld A. and Kak A.C. 1976: Digital image processing. Academic Press, New York, 455 pp.

Rouse J.W., Haas R.H., Schell J.A. and Deering D.W. 1973: Monitoring vegetation systems in the Great Plains with ERTS. Third ERTS-1 Symposium, Goddard Space Flight

Center, NASA SP-351, pp 309-317.

Sabins F.F. 1978: Remote sensing, principles and interpretation. W.H. Freeman and Company, San Francisco, 426 pp.

Saebø H.V., Braten K., Hjort N.L., Llewellyn B. and Mohn E. 1985: Contextual classification of remotely sensed data: statistical methods and development of a system. Norwegian Computing Center, report no 768, 300 pp.

Saxon E.C. 1984: Multitemporal texture transformed Landsat imagery for mapping ecological gradients. Proceedings of the 3rd Australasian Remote Sensing Conference, Queensland, pp 255-259.

Schowengerdt R.A. 1983: Techniques for image processing and classification in remote sensing. Academic Press, London, 249 pp.

Shih E.H. and Schowengerdt R.A. 1983: Classification of arid geomorphic surfaces using Landsat spectral and textural features. Photogrammetric Engineering and Remote Sensing 49, no 3, pp 337-347.

Siegal B.S. and Goetz A.F.H. 1977: Effects of vegetation on rock and soil type discrimination. Photogrammetric Engineering 43, pp 191-196.

Sinclair D. 1984: The census: It can be done more accurately with spaceage technology. Photogrammteric Engineering and Remote Sensing 50, no 1, pp 80.

Singh A. 1984: Some clarifications about the pairwise divergence measure in remote sensing. Int Journal of Remote Sensing 5, pp 623-627.

Smedes H.W., Hulstrom R.L. and Ranson K.J. 1975: The mixture problem in computer mapping of terrain: improved techniques for establishing spectral signatures, atmospheric path radiance and transmittance. NASA Earth Resources Survey Symposia, Houston, pp 1099-1159.

Som R.K. 1973: Recall lapse in demographic enquiries. Asia Publishing House, New York, 212 pp.

Stern M. 1979: Ett försök till utvärdering av Landsat 1 digitala och analoga data för kartering av vegetation och markanvändning i ett skånskt testområde. Mimeo, Department of Physical Geography, University of Lund, 22 pp.

Stern M. 1981: Landsat data for monitoring rural settlement and population - a test in the Umm Ruwaba region, the Sudan. Proceedings of the Int. Symp. on Remote Sensing of Arid and Semi-arid Lands, Cairo, Egypt, pp 437-447.

Stern M. 1984: Landsat data for population estimates - Approaches to inter-censal counts in the rural Sudan. COSPAR, Adv Space Research 4, pp 69-73.

Strahler A.H. 1980: The use of prior probabilities in maximum likelihood classification of remotely sensed data. Remote Sensing of Environment 10, pp 135-163.

Strahler A.H. and Woodcock C.E. 1984: Image variance and spatial structure in remotely sensed data. Third Australasian Remote Sensing Conference, Queensland, pp 512-518.

Swain P.H. and Davis S.M. 1978: Remote sensing - The quantitative approach. McGraw-Hill Int Book Company, 396 pp.

Taranik J.V. 1978: Principles of computer processing of Landsat data for geologic applications. US Geological Survey, Open-File report 78-117.

Tewfik M. 1977: A pilot project for the Khors, Eastern Sudan. Ekistics 258, pp 268-276.

Townshend J.R.G. 1981: Image analysis and interpretation for land resources survey. In Townshend (ed):

Terrain analysis and remote sensing. George Allen & Unwin, London, 232 pp.

Townshend J.R.G. 1981b: The spatial resolving power of earth resources satellites. Progress in Physical Geography, pp 32-55.

Tucker C.J. 1979: Red and photographic infrared linear combinations for monitoring vegetation. Remote Sensing of Environment 8, pp 127-150.

United Nations 1977: Desertification: Its causes and consequences. Pergamon Press, 448 pp.

Verdin J. 1983: Corrected vs. uncorrected Landsat 4 MSS data. Landsat Data User's Note 27, pp 4-8.

Watson R.M. and Tippet C.I. 1975: Sudan pilot livestock census, South Kordofan, Volume 2. Khartoum: Ministry of Agriculture, Food & Natural Resources, 77 pp.

Watson R.M., Tippett C.I., Rizk F. and Jolly F. 1976: Sudan national livestock census and resource inventory, Volume 28. Khartoum: Veterinary Research Administration, 40 pp.

Welch R. 1980: Monitoring urban population and energy utilization patterns from satellite data. Remote Sensing of Environment 9, pp 1-9.

Wiegand C.L., Gausman H.W., Cuellar J.A., Gerbermann A.H. and Richardson A.J. 1974: Vegetation density as deduced from ERTS-1 MSS response. Third ERTS Symposium, NASA SP-351, pp 93-116.

Wellar B.S. 1969: The role of space photography in urban and transportation data series. Proceedings of the Sixth Int Symp on Remote Sensing of Environment, pp 831-854.

Weszka J.S., Dyer C.R. and Rosenfeld A. 1976: A comparative study of texture measures for terrain classification. IEEE Transactions on Systems, Man, and Cybernetics, vol SMC-6, no 4, pp 269-285.

Work Jr E.A. and Gilmer D.S. 1976: Utilization of satelite data for inventorying prairie ponds and lakes. Photogrammetric Engineering and Remote Sensing 42, pp 685-694.